PENGUIN BOOKS

A BRIEF HISTORY OF INFINITY

Paolo Zellini was born in Trieste in 1946 and graduated with a degree in Mathematics at the University of Rome in 1970. He is now a professor at the University of Rome and at the CNR (the Italian National Council for Scientific Research).

David Marsh is Professor of Italian at Rutgers University and a specialist in Renaissance studies. His books include *The Quattro-cento Dialogue* and *Lucian and the Latins*, and he has edited and translated works by Francesco Petrarca, Leon Battista Alberti, Leonardo da Vinci and Giambattista Vico.

CW00550849

PAOLO ZELLINI

A Brief History of Infinity

Translated by DAVID MARSH

PENGUIN BOOKS

PENGUIN BOOKS

Published by the Penguin Group
Penguin Books Ltd, 80 Strand, London WC2R ORL, England
Penguin Group (USA) Inc., 375 Hudson Street, New York, New York 10014, USA
Penguin Group (Canada), 10 Alcorn Avenue, Toronto, Ontario, Canada M4V 3B2
(a division of Pearson Penguin Canada Inc.)
Penguin Ireland, 25 St Stephen's Green, Dublin 2, Ireland
(a division of Penguin Books Ltd)
Penguin Group (Australia), 250 Camberwell Road, Camberwell, Victoria 3124, Australia
(a division of Pearson Australia Group Pty Ltd)
Penguin Books India Pvt Ltd, 11 Community Centre, Panchsheel Park, New Delhi – 110 017, India
Penguin Group (NZ), cnr Airborne and Rosedale Roads, Albany, Auckland 1310, New Zealand
(a division of Pearson New Zealand Ltd)
Penguin Books (South Africa) (Pty) Ltd, 24 Sturdee Avenue, Rosebank 2196, South Africa

Penguin Books Ltd, Registered Offices: 80 Strand, London WC2R ORL, England

www.penguin.com

Breve storia dell'infinito first published in Italy by Adelphi Edizione 1980
This translation first published by Allen Lane 2004
Published in Penguin Books 2005
3

Copyright © Adelphi Edizione, 1980
Translation copyright © David Marsh, 2004
All rights reserved

The moral right of the translator has been asserted

Set by Rowland Phototypesetting Ltd, Bury St Edmunds, Suffolk
Printed in England by Clays Ltd, St Ives plc

Except in the United States of America, this book is sold subject
to the condition that it shall not, by way of trade or otherwise, be lent,
re-sold, hired out, or otherwise circulated without the publisher's
prior consent in any form of binding or cover other than that in
which it is published and without a similar condition including this
condition being imposed on the subsequent purchaser

Contents

Translator's Note

Wherever possible, I have quoted the English originals of texts cited by the author, and cited the standard English translations of literary and philosophical texts, occasionally modifying some passages to suit the context. Classical authors generally follow the versions of the Loeb Classical Library. For help on specific chapters, I owe special thanks to Carlo Lancellotti, Samuel Levey and Tim Maudlin. Paolo Zellini generously read the entire manuscript, correcting many of my mistakes and occasionally adding his own clarifications.

David Marsh
May 2002

I

Aristotle's *Apeiron*: Limit and the Unlimited

'There is one concept that corrupts and confuses the others. I am not speaking of Evil, whose limited sphere is Ethics; I am speaking of the Infinite.' Thus Jorge Luis Borges introduces his brief survey of the Infinite in *Otras Inquisiciones*.[1] Not only here, but elsewhere, his conception of infinity, often disguised in related notions, emerges as an absolute metaphysical evil that operates in the cosmos as a seed of disorder and absurdity. There is nothing more dangerous than the loss of limits and measure. This is the error caused by the infinite: we lose sight of the meaning implicit in the relative perfection of what is concretely determined and formally complete, and so are led astray into the void or into a labyrinth with no exit.

In retracing the origins of such notions of infinity, it might be auspicious to begin with a statement by Immanuel Kant. 'In the great wealth of our languages,' he writes, 'the thinking mind nevertheless often finds itself at a loss for an expression that exactly suits its concept, and lacking this it is unable to make itself intelligible either to itself or to others. Coining new words is a presumption to legislate in language that rarely succeeds, and before we have recourse to this dubious means it is advisable to look in a dead, learned language for an expression that might be suitable to formulate this concept.'[2]

If we seek to find a metaphysical justification for the *horror infiniti*

---

1. [See 'Avatars of the Tortoise', in Borges, *Other Inquisitions 1937–1952*, trans. Ruth L. C. Sims (Austin, Texas, 1964), pp. 109–15. Translator's Note.]
2. Immanuel Kant, *Critique of Pure Reason*; Italian translation (Milan, 1976), p. 373. ['Transcendental Dialectic, First Book, First Section', A 312, B 369: English version by P. Guyer and A. W. Wood (Cambridge, 1998), p. 395. Translator's Note.]

and the labyrinthine fascination of infinity in Borges' literary experiments, we might employ a Greek word, which would align our thoughts and reflections with a viewpoint that can be inferred from the metaphysics of Anaximander, the Pythagoreans and Aristotle. In Greek thought, references to infinity meant resorting to a term whose meaning is clearly not synonymous with the connotations of our word 'infinite'. Their term for this was *apeiron*, which literally means 'without limits' ('limit' in Greek is *peras*) and hence 'unlimited'. The range of meanings traditionally suggested by the term *apeiron* clearly makes Aristotle's assertions seem plausible: for these meanings reveal that the infinite, by its nature, is not only divine and incorruptible, but is also ambiguous and defies analogies and analysis. Most important of all, Aristotle did not hesitate to assign the status of principle to the infinite. In his *Physics* 3.2.203b6 we read: 'Everything is either a principle or derived from a principle. But there cannot be a principle *of* the infinite, for that would be a limit of it. Further, it is both uncreatable and indestructible, since it is a principle. For there must be a point at which what has come to be reaches its end, and also a termination of all passing away. That is why, as we say, there is no principle *of this*, but it is this which is held to be a principle of all other things, and to encompass all and to steer all, as those assert who do not recognize other causes than the infinite . . .' And in *Physics* 3.4.203b10 Aristotle continues: 'Further, it is different, for it is deathless and imperishable as Anaximander says, with the majority of the physicists.'

Are there proofs of the existence of the infinite? Aristotle lists a few of them: time (which is manifestly infinite), and the division of magnitudes. But there are even more profound reasons that prevent us from reducing the infinite to pure imagination: 'If coming to be and passing away are not exhausted, it is only because that from which things come to be is infinite,' Aristotle argues. 'Again, the limited always finds its limit in something, so that there must be no limit, if everything is always limited by something different from itself. Above all, the principal reason which produces a difficulty common to everyone is that, since they are never fully exhausted in our thought, mathematical numbers and magnitudes and everything that is beyond the heavens seem infinite.'[3]

---

3. Aristotle, *Physics* 3.4.203b20.

Hence, the difficulty inherent in infinity lies in its inexhaustibility. What is infinite (Aristotle cites the example of the set of numbers) can never be present in our thought in its totality. This property of the infinite – meaning *apeiron* – is so characteristic that it constitutes a first definition of it. In other words, we may assert that any set of objects is unlimited when, attempting to specify all its elements singly, we cannot constitute a whole, because in every case there will always be some element that we have not yet considered. For example, if we consider whole numbers, they form an infinite set, because we may count an arbitrarily large number of them without ever reaching a limit beyond which there are no other numbers as yet uncounted. Aristotle writes: 'The infinite [*apeiron*] is not what has *nothing* outside of it, but what always has *something* outside of it.'[4] The unlimited cannot, therefore, in any case be considered as a complete totality: for whatever is complete has a boundary, and a boundary is a limiting element, whereas the *apeiron*, by its intrinsic meaning, signifies the absence of any limit.

Apeiron is indissolubly linked to a negative notion: the expression of its incompleteness, and of its unrealized and unrealizable potential. For this reason, the term 'indefinite' or 'unlimited' seems preferable to 'infinite' if we seek one word that combines the varied meanings associated with *apeiron*. Indeed, our word 'infinite' often refers to an idea of perfection which is alien to the meaning of *apeiron*, and may therefore conceal the fact that we are tempted to reduce the incompleteness of the Greek 'unlimited' to a reality foreign to it. This distinction is important, especially because it suggests the ultimate reasons behind solutions to mathematical and geometrical problems involving the infinite. The refusal of Greek mathematicians (with a few exceptions) to introduce a real infinity can only be understood if we bear in mind the negative concept implied by *apeiron*.

The character of non-existence implicit in *apeiron* is suggested by its analogy to *steresis*, or privation, which constitutes a necessary if momentary premise for all movement towards evolution.[5] Thus, coming into being appears at every moment to be a synthesis of limit

---

4. Aristotle, *Physics* 3.6.207a1.
5. Cf. Aristotle, *Physics* 3.7.207b35–208a1.

(*peras*) and of the unlimited (*apeiron*). Limit is what makes every object exist concretely, by constantly endowing it with its proper form and individuality. It is also what determines the logical order of events, by removing them as far as possible from pure chance. On the other hand, neither history nor evolution would exist if, in addition to limit, there did not also exist a principle of an opposite nature, which thwarts the tendency of every object to remain rigidly fixed within the contours imposed on it by the principle of limit. The unlimited is precisely this sort of principle. It appears to be a negative and destructive principle. For disrupting the order imposed by limit clearly means bringing reality back to an amorphous and disorganized state, in which each thing loses its recognizable form as a concrete entity, and in which events appear disconnected, unpredictable, and liable to evolve irrationally. Yet this state is a necessary premise for the operation of limit, which at every moment corrects the situation of indefinite potentiality implicit in the unlimited, and imposes a rational development on events.

Typically, the principle of *apeiron* seems to act in the sphere of Becoming, where it manifests itself both in the dissolution of forms and as an element of chance.[6] (According to Anaximander, it is *apeiron* itself whose primal movement triggers this Becoming.) And conversely, the idea of Becoming helps explain the existence of an *unlimited* set. For the elements of such a set do not exist simultaneously: that is, they are not actually present singly. Rather, they only exist as a kind of historical series: that is, one *after* the other, in an unending sequence, exactly as 1 is followed by 2, 2 by 3, and so forth. In this sense, Aristotle regards the existence of the infinite as potential rather than actual. As such, it is more a parallel to the material principle of existence than to its formal principle; indeed, we may say it is the antithesis of the formal principle.[7]

Nevertheless, *apeiron* is a 'divine, deathless and indestructible principle,' as Aristotle, like Anaximander, seems to maintain. But can a negative principle be associated with the divine? Anaximander uses the

---

6. On the relation between chance and material (unlimited) principle, see Gaston Milhaud, 'Le hasard chez Aristote et chez Cournot', *Revue de Métaphysique et de Morale* 10 (1902): 667–81.

7. Aristotle, *Physics* 3.6.207a21.

term 'infinite' in two different senses. The first is the temporal sense related to the inexhaustible series of cosmic cycles, and the second is that related to the permanence and timelessness of their final substratum. In the latter sense, Anaximander uses *apeiron* as a synonym of the divine (*to theion*); but it has been observed that, unlike *theos* (God), *to theion* refers to a neutral metaphysical principle which is quite compatible with the idea of pure negativity denoted by *apeiron*.[8] God can only be defined as an undefinable being; and negative theology, by denying all of God's attributes, provides the only possible description of him. This is precisely how the 'divine' would be referred to by Pseudo-Dionysius, by John Scotus Eriugena in *On the Division of Nature*, and by Nicholas of Cusa. The latter says that Plato himself, according to Proclus, spoke of the first principle by denying it any attribute.[9] And Marsilio Ficino attests that Orphic thinkers called the divine mind *apeiron omma*, 'the infinite eye'.[10]

Paul Tannery may have been right in deriving Anaximander's *apeiron* not from *peras* 'limit', but from *peira* 'knowledge or experience', and thus in calling it the 'unknowable' or 'unfathomable'.[11] (Plato often uses *apeiria* in the sense of 'inexperience'.) But the two meanings are similar and complementary. Between the indefinite One and the finite creature, there lies an insuperable distance: a potential infinite, which can be represented as an incalculable number of successive steps that, traversed one at a time, never reach their goal. (Homer writes that Zeus is separated from the earth 'by the unbridgeable ether': *di'aitheros atrugetoio*.[12]) According to Aristotle too, something which has no limit (*peras*) cannot be exhaustively represented in our thought, and is therefore unknowable.

For Aristotle, the unlimited was the material substratum of visible

8. G. B. Burch, 'Anaximander, the First Metaphysician', *Review of Metaphysics* 3 (1949): 137–60.

9. Nicholas of Cusa, *De beryllo* 11, in *Opere filosofiche* (Turin, 1972), p. 650.

10. See Giordano Bruno, *De la causa, principio et uno*, ed. G. Aquilecchia (Turin, 1973), p. 68.

11. *Revue de Philosophie* 4 (1904): 703–7.

12. Cf. Roberto Mondolfo, *L'infinito nel pensiero dell'antichità classica* (Florence, 1956), p. 34. Mondolfo variously translates *atrugetos* as 'inexhaustible', 'infinite' and 'uncrossable'.

objects, which could in some way be conceived of as their indefinite divisibility.[13] In fact, a corporal entity is in itself a complete form, an unrepeatable 'whole' constructed by the formal, limiting principle on the substratum of infinite potentiality, of the infinitesimals that make up the whole. When it comes to dividing an entity, we begin to speak of its 'parts' rather than the 'whole', and its form already seems to be compromised. But when division is ideally carried to its ultimate consequences, that is, to infinitesimals, the original form appears fragmented, unrecognizable, and ultimately transformable into something else.[14] Infinite division implies the tendency to return to pure potentiality, to the substantial principle (*hyle*) of all subsequent existence. In Anaximander's *apeiron* too we find the ultimate principle of materiality and the whole cause of the birth and destruction of all things. From *apeiron* the heavens are born through separation, as in general are all worlds (which are infinite) in a cyclical alternation of birth and destruction.

Hence, the Greeks' Chaos and Night are mythical synonyms of *apeiron* understood as an ultimate substantial principle, as are the Babylonian Tiâmat and the Hindu Mṛtyu.[15] Guilty of conspiring with Evil, Tiâmat is banished by the god Marduk to the far corners of the universe; and Mṛtyu, the goddess of death and hunger, suffers the same fate at the hands of the Devas. But while banished to the ends of the World, the powers of Nothing and of the Indefinite continue to play an essential role in encompassing and sustaining all things (Anaximander's *apeiron* is the all-encompassing *to periechon*); and at the end of every cosmic cycle everything returns to them.

Thus, myth reveals the recondite and enigmatic nature of the infinite: originally absolute and neutral, it changes as form is created into the negative pole of its own illusory duality. Plotinus writes as follows: 'The unlimited is dual. And what is the difference between the two

---

13. Aristotle, *Physics* 3.6.207a22.

14. At the same time, Aristotle denies that the alternation of coming to be and passing away implies an actual infinite. While breaking up means returning to the unlimited, it is possible that the passage from passing away to coming to be takes place without infringing the boundaries of limit (*Physics* 3.8.208a9).

15. Cf. Mondolfo, *L'infinito*, pp. 56–7, and the *Bṛhad-āraṇyaka-upaniṣad*.

unlimiteds? They differ as the archetype differs from the image. Is the unlimited here, then, less unlimited? More, rather; for in so far as it is an image which has escaped from being and truth, it is more unlimited. For unlimitedness is present in a higher degree in that which is less defined; and the less in the good means more in the bad.'[16]

Parmenides and Aristotle quite often use negative terms such as *ageneton, anolethron, ateleston, apeiron, aphtharton*: 'uncreated, inde-structible, endless, infinite, undecaying'. Such words insistently point to an ineffable reality which we can only try to verbalize at the risk of losing ourselves in labyrinths of the false infinite – in other words, of an illusory image of the totality we seek to intuit. Damascius writes that Plato, having attained the One, fell silent. 'Indeed, as soon as he began to discourse about the absolutely non-existent, he drew back to avoid falling into an ocean of disparity or, better, into a void with no foothold.'[17]

We read in the *Tao te Ching* that 'the alternation of Non-being and Being produces the marvel of the one and the boundaries of the other.'[18] When reality is regulated by boundaries, the true infinite can only become manifest by being condensed into a finite form. For anyone who remains in the realm of Being, the transgression of limits and the desire for miracles are always close to being confused with the pure and simple disintegration of form. This constitutes the perennial similarity between holiness and madness, and is also one of the principal and most profound reasons for the attraction of evil. 'A certain fellow,' Kafka relates, 'was amazed at how easy it was to travel the road of eternity; in fact he was coasting along it downhill.'[19]

According to Theophrastus, Anaxagoras conceived the infinite in the same way as Anaximander. Anaxagoras called 'unlimited' (*apeiron*) the primordial mixture of chaos, in which nothing existed because forms had not yet been conceived. According to his Fragment 1, preserved in Simplicius' *Physics*, he described how 'all things were

---

16. Plotinus, *Enneads* 2.4; Italian translation (Bari, 1973), vol. 1, p. 196.
17. Damascius, *Quaestiones de primis principiis* (Frankfurt, 1826), cited by Raniero Gnoli in his introduction to Nāgārjuna, *Madhyamaka kārikā* (Turin, 1968).
18. *Tao te Ching*; Italian translation (Milan, 1973), p. 27.
19. Franz Kafka, *Confessioni e immagini* (Verona, 1960), p. 62.

together, unlimited in quantity and smallness: for the small too was unlimited'. Anaxagoras described this extreme rarefaction – implied by reducing everything to infinitesimal particles lacking any organizational principle – by saying that 'all things were pervaded by air and ether, which were both unlimited'. But even chaos conceals a powerful elementary virtue, as Nicholas of Cusa writes in *De ludo globi* (*The Game of Spheres*). The elementary particles, which Anaxagoras called 'homoeomeries', can be set in motion by the ordering intervention of Intellect, which dissolves their mutual indistinguishability and causes forms to emerge through differentiation. Yet unlike the atoms of Democritus, the homoeomeries of Anaxagoras are not indivisible particles that limit the divisibility of bodies and prevent their infinite fragmentation. Rather, they are themselves infinitely divisible into infinite elements. 'In fact,' Anaxagoras wrote (Fragment 3), 'there is no limit to what is small, for there is always something smaller.' And by the same token, 'of the large there is always something larger. In terms of quantity, the large is equal to the small; but in relation to itself, each thing is both large and small.'

Yet somewhere within this inexhaustibility of the unlimited, there are causes for the differentiation of forms. If we divide a body indefinitely, it becomes impossible to arrive at its elementary and indivisible constituents. But in the course of this division we may encounter, at a nearly infinitesimal level of elements, some precise ratio between them, or some imperceptible difference in distribution, that favours the generation of one form over another. The possibilities of diversity are infinite, and many of them depend on small or even infinitesimal variations in the invisible proportions and interrelations that exist between their constituent elements. Aristotle sought to prove that small alterations are the cause of great alterations, not by virtue of themselves, but because their principles have different tendencies. And these principles, Aristotle added, while small in size, are great in potency, and from them even opposites, such as male and female, may arise.[20]

Thus, already in Anaxagoras we find the principal themes that recur in all later analyses of the infinite. For example, the incalculable number

---

20. Aristotle, *On the Generation of Animals* 5.7.788a13.

of combinations between infinitesimals, and the incomprehensible results generated by countless formal differences, clearly hint at the existence of a continuity principle operating in nature, which Leibniz was to make a central point of his own speculations.[21] The inexhaustibility of the unlimited, and the impossibility of finding an absolute minimum or maximum, not only became a central object of continual reflection – from the philosophical schools of Oxford or Paris in the fourteenth century, to the metaphysics of many Renaissance thinkers, and culminating in Leibniz and Newton – but also formed the basis for the definite formulation of infinitesimal calculus arrived at by Cauchy and Weierstrass in the nineteenth century.

Still, the most explicit declaration of the fundamental bipolarity that governs the movement of the cosmos – expressed in terms of 'finite' and 'indefinite' – came from the Pythagoreans. Hippolytus records the words of the Chaldean Zarata (Zarathustra) to Pythagoras: 'From the beginning there have been two causes for the things that exist, the father and the mother. The father is light, and the mother darkness. The qualities of light are warmth, dryness, lightness and speed; and those of darkness are cold, wetness, weight and slowness. The whole cosmos is composed of these two things, male and female.'[22] According to Hippolytus, the two principles to which Chaldean wisdom gave these names were later called 'limit' (*peras*) and 'the unlimited' (*apeiron*) by the Pythagorean Philolaus, who associated limit with good and the unlimited with evil.

The universal significance of this Pythagorean doctrine lies in the fact that we can intuit concretely the antithetical and dialectical interaction of these two principles everywhere. In his studies of nature, Goethe would implicitly rediscover this for the modern age; and his rich intuition of its potential applicability to classifying the real world may justly remind us of the Pythagoreans. Goethe used the criteria of polarity and of rhythmic oscillation as primary research *methods* for the fundamental classification of phenomena. In the alternation of

---

21. Cf. Mondolfo, *L'infinito*.

22. Hippolytus, *Refutatio contra omnes haereses*, ed. Wendland, 1–2, 12, p. 7, cited in *I Presocratici: Testimonianze e frammenti*, 2 vols. (Bari, 1969), 1:123.

systole and diastole, he perceived the essence of all organic development, as in the typical case of a plant's growth (*Morphology of Plants*, 73, 121). In fact, this consists of a continual alternation of (unlimited) expansions and (limiting) contractions. Starting from the seed, which has the maximum concentration (limit), a plant arrives at its maximum development in the leafy stalk (unlimited); then the calyx emerges by contraction (limit) and the petals by a new expansion (unlimited). The movement of expansion culminates in the fruit, in which the entire stalk is concealed in a sort of condensed form.

Goethe's work was also cited by Hermann Broch as a prototype of how science and literature could be reunited since both fields were permeated by the infinite. More than anyone else since the Renaissance, Broch wrote, Goethe was able to fuse scientific interests with artistic goals, by assigning to poetry the power of symbolically evoking the *maternal forces* of the cosmos – a common root of all polarity (an infinite that precedes all successive comparison between the limit and the unlimited) towards which the aspirations of science were also directed.

Limit and the unlimited were also the principles from which all of Pythagorean arithmetic developed, as we are told by Plutarch, Theon of Smyrna, Nicomachus of Gerasa and Boethius. In this arithmetic, the fundamental categories of integers are divided according to analogous bipolarities – odd and even, or equal and unequal, or divisible and indivisible. These polarities also govern the successive unfolding of natural numbers and of proportions in the schemes resembling 'lambda' and 'delta', which are common to both the Platonic and Pythagorean traditions, and which were reinterpreted in the nineteenth century by Albert Freiherr von Thimus in *Die harmonikale Symbolik des Alterthums* (*Harmonic Symbolism in Antiquity*) (Cologne, 1868).

Moreover, such arithmetic procedures further allow us to deduce a function of limit which in a certain sense lies 'outside' any polarity: this function suggests that limit and the unlimited can be reunited in an entity which, while being perhaps definable, is inextricably bound up with non-being. This happens whenever successive definitions of whole numbers using proportions describe a progressive approach to some irrational number. For example, from the mathematics of Nicomachus of Gerasa we deduce a method for approximating the

Golden Number, $(1 + \sqrt{5})/2$, by means of successive ratios approaching it from above and from below – a process that indefinitely extends the large–small polarity denoted by the Greek *mega kai mikron*.[23]

The fundamental laws of arithmetic also hold good in geometry. In his *Commentary on Book One of Euclid's Elements*, Proclus recognized that all of geometry can be divided between limit and the unlimited, as exemplified in the bipolarity of the straight line and the curve. This truth persists in mathematical speculation from the Renaissance to Leibniz, who would likewise derive the composition of every geometric figure from what is rectilinear and circular.

As Paul Feyerabend shows in his book *Against Method*, Copernicus too had a similar concept in mind when he observed that the movement of a falling rock must be a 'combination of the rectilinear and the circular'.

According to Simone Weil, limit is always exceeded, but it responds by imposing a compensatory oscillation, an indefinite process which is represented by circular motion.[24] In Aristotle's *Problems*, the Pythagorean Archytas, when asked why the trunk and branches of a tree, or the legs and arms of a man, are round, is said to have answered: 'There exists a natural movement tending towards the identical which, by observing the law of limit, returns on itself, thus giving rise to circles and other round figures.'[25]

As Kant for one pointed out, the unlimited manifests itself psychologically in the continuous referring of every predicate to a corresponding subject. This referring is indefinite because one never arrives at an ultimate subject, just as one never arrives at the ultimate substratum of every accident. It would seem that self-consciousness, meaning the immediate perception of oneself as thinking subject, would provide a firm basis for an infinite regression of predicates and subjects. But this is purely an illusion, Kant says; there can be no true knowledge of an absolute proper being precisely because it is impossible to give it a predicate. In this sense, discursive thought is intrinsically unlimited; it

23. Cf. Nicomachus of Gerasa, *Introduction arithmétique* (Paris, 1978), p. 32.

24. Simone Weil, *Cahiers*, vol. 2 (Paris, 1972), p. 32. [*The Notebooks of Simone Weil*, trans. Arthur Wills, 2 vols. (London, 1956), p. 162. Translator's Note.]

25. Aristotle, *Problems* 16.9.915a25, cited in *I Presocratici* 1:490.

runs through an infinite potential which has no term and no solution. In this process, limit coincides with the step-by-step application of the identity principle ('*A* is *A*' or '*x* = *x*') and with our tendency to enclose every cognitive object within a sort of short circuit that removes it from discursive flux and isolates it from everything from which it differs.

In his *Philebus* 16D, Plato too asserted that the opposition between finite and infinite is an intrinsic feature of our thought and discourse which never ends or grows old. Plato reiterates the view that everything has ingrained within it both limit and the unlimited or, in equivalent terms, the one and the many. In the division of bodies, we see another clear manifestation of the irreducible conflict between unity and multiplicity. (And we must bear in mind that unity and multiplicity are synonyms of limit and the unlimited.) Each object is recognizable as a complete whole, synthetically visible and tangible (or conceivable). But it is also the aggregate of all its parts, which constitute an infinite multiplicity that defies any actually exhaustive enumeration. So can the one be reduced to the many? And, vice versa, can the many be reduced to the one? According to Plato, the problem can be concretely resolved, and indeed is, by the actual synthesis or mixture of limit and the unlimited in all things. The human body, for example, is entirely composed of parts; and these parts respect well-defined numerical and harmonic proportions. In these, the unity of the whole is articulated by measure; and rather than breaking up into the unlimited, it remains in an intermediate zone where rhythm and beauty are concretely developed and revealed. Thus, number, *arithmos*, which is synonymous with measure and harmony, is a sort of pause or point of mediation between limit and the unlimited. It is a Promethean gift – Aeschylus called Prometheus the 'father of number' – which ensures the possibility of a stable and ordered existence for mankind, for it rescues us from a precarious balance between absolute unity and absolute multiplicity. Such a balance is out of the question, since any direct approach to the absolute may ultimately be reduced to a dangerous encounter with the false infinite.

In antiquity, the *horror infiniti*, or dread of the infinite, reached its culmination in the writings of Boethius. He went so far as to speak of the unlimited as a 'monster of malice' (*malitiae dedecus*) which rests on no principle and perennially defies any sort of definition: an object that both science and philosophy gladly repudiate, since they recognize its resistance to any attempt at comprehension.[26] The principal quandary, already expressed by Aristotle, was that the unlimited contradicted the law that, logically and naturally, act precedes potential. The most typical exception, Boethius explains, is number.[27] For while it can increase limitlessly, and be represented at every step as an actual entity, such as 10 or 100, number never reaches the conclusion towards which it seems directed by its indefinite capacity to increase. The act that should encompass and give meaning to its potential appears not to exist at all.

In the modern era, Spinoza, Hegel and Leopardi grasped the negativity of the potential infinite and related it to desire and the imagination.

Spinoza called the false infinite the 'infinite of the imagination', and Leopardi wrote that the infinite is 'the offspring of our imagination, of our smallness and at the same time of our pride . . . It is a dream, not a reality, because we have no proof of it, not even by analogy.'[28]

Leopardi also distinguished the infinite from the indefinite, and saw in the latter a seductive counterfeit of the former, a deceptive product produced when our imagination strives to intuit totality. He wrote: 'Not only the cognitive faculty, or the faculty of loving, but even the imaginative faculty is incapable of the infinite, or of conceiving infinitely, and is only capable of the indefinite, and of conceiving indefinitely. This delights us because the soul, seeing no boundaries, receives the impression of a sort of infinity, and confuses the indefinite with the infinite, yet without effectively comprehending or conceiving any infinity. Rather, in the most vague and indefinite imaginings, which are therefore the most sublime and delightful, the soul expressly feels a certain anguish, a certain difficulty, a certain insufficient desire, and a decided inability to encompass the entire measure of its imagining, conception, or idea.'[29]

---

26. Boethius, *De institutione arithmetica*, ed. Friedlin (Leipzig, 1867), pp. 9, 126.
27. Boethius, *Commentarii in librum Aristotelis* (Leipzig, 1877), pp. 206–7.
28. Giacomo Leopardi, *Zibaldone* 4177–8, in Leopardi, *Opere*, 2 vols. (Florence, 1976), 2:1099.
29. Leopardi, *Zibaldone* 472–3, in *Opere* 2:166.

Nothing so perfectly reveals the illusory quality of the unlimited as the fact that an authentic object of desire is indefinable. In his *Zibaldone* 165-7, Leopardi wrote that the most intimate reason for the human mind's propensity towards the unlimited lies in its desire for pleasure and in the resistance of this desire to any definitive satisfaction. The deep-rooted nature of desire lies in its projection on to an absolute and non-existent object, which is not identifiable with any definite pleasure. For this reason, its nature 'materially involves infinity, because each single pleasure is circumscribed, but not pleasure in general, with its indeterminate extension. Loving pleasure in general, the soul embraces the whole imaginable extension of this feeling without being able to conceive it, since no clear idea can be formed of a thing that the soul desires to be unlimited.'[30] As a result, the attainment of any desired object, which is limited and circumscribed, creates a sort of vacuum in the soul, an anticipatory pain on our discovering its illusory nature and foreseeing its end. Consequently, the infinite pleasure that we cannot attain in reality finds its ideal place in the imagination, which is the source of the fictitious character of all our hopes and illusions.

'The imagination,' Simone Weil wrote, 'is always linked with desire . . . Only desire without object is empty of imagination. Beauty is naked, unshrouded by imagination. God's real presence is in everything that is unshrouded by imagination. The consecration of the Host is a supernatural operation which reduces a piece of matter to nakedness. Beauty seizes upon the finality in us and empties it of all ulterior end; seizes upon desire and empties it of all ulterior object, by presenting it with an object actually present and thus preventing it from launching out toward the future.'[31]

The torment of Tantalus, which Hegel in his *Aesthetics* viewed as a

---

30. Leopardi, *Zibaldone* 165, in *Opere* 2:80.
31. Simone Weil, *Cahiers*, vol. 3 (Paris, 1974), p. 192. [*Notebooks*, trans. Wills, p. 553. Translator's Note.] In the *Bhagavad Gita* (2.47) we read that man is called to act, but not to enjoy the fruit of his actions. Hence the injunction (16.23) 'Never consider the fruit of your action as a motive,' because 'whoever . . . acts under the sway of desire obtains neither perfection nor happiness nor the ultimate goal'; Italian translation (Milan, 1976), pp. 39, 156.

symbol of the false infinite, symbolizes the indefinite repetition of an act that strives to attain an illusory goal. Monotony is always present as an element in the potential unlimited. The punishments of Tantalus and Sisyphus, Hegel writes, are 'the inherently measureless, the bad infinite, the longing for the necessity of being, or rather the insatiability of natural subjective desires which, in their constant repetitiveness, never reach the ultimate calm of fulfilment. For the Greeks, the exact sense of the Divine, unlike modern yearning, did not regard egress into the boundless and vague as what was supreme for man; the Greeks regarded it as damnation and relegated it to Tartarus.'[32]

All the aspects of the 'positive' sense of limit, and of the disintegrating force of the unlimited, were also intuited by Robert Musil. In his novel *The Man without Qualities*, he describes the 'positive' value of boundaries and measure in the figure of Arnheim. This character possesses an awareness of a profound moral secret, an innate sense which has largely been lost today: the simple intuition that we are not allowed to do everything. 'All discipline, abstinence, chivalry, music, morality, poetry, form, taboo, had no deeper purpose than to give the correct limits, a definite shape, to life.' 'There is no such thing as boundless happiness,' Musil writes. 'There is no great happiness without great taboos. Even in business to pursue one's advantage at all costs is to risk getting nowhere. Keeping within one's limits is the secret of all phenomena, of power, happiness, faith, and the key to the task of maintaining oneself as a tiny human creature within the boundless universe.'[33]

But limit embodied in some moral rule or norm cannot constitute an absolute. In order to survive, it must be nourished by its own unceasing energy and by its intrinsic necessity, for it is always subject to the disintegrating effect of the unlimited. Hence it is just as true, as Musil observes, that 'our morality is the crystallization of an inner

---

32. Hegel, *Ästhetik* II.2.1; Italian translation (Turin, 1967), pp. 524–5. [English version, *Aesthetics*, trans. T. M. Knox, 2 vols. (Oxford, 1975), vol. 1, p. 466. Translator's Note.]

33. Musil, *Der Mann ohne Eigenschaften*, I.105; Italian translation, *L'uomo senza qualità* (Turin, 1962), p. 488. [English version, *The Man Without Qualities*, trans. Sophie Wilkins (New York, 1995), p. 548. Translator's Note.]

movement that is completely different from it.'[34] 'The power for good present in us instantly eats its way through the walls if it gets locked into solid form, and immediately uses that as a bolt hole to evil! . . . All emotions refuse to be chained.' And again: 'Faith mustn't ever be more than an hour old!'[35]

Whenever the crisis of an epoch threatens to destroy the prevalent morality and to eliminate decayed and empty forms, there is one ultimate defence that permits our ethical survival. We see this in Ulrich, the hero of *The Man Without Qualities*. For him, if a morality is to possess real growth power and to withstand recurrent catastrophes, it must not be founded on an order established for all time, but on the unceasing activity of creative imagination. Rather than being governed arbitrarily, such imagination reshapes the varied inspirations furnished by the infinite complex of the possibilities of life. In this way, morality should be built on the successive steps of our experience; it should not be some eternally ordained ideal that man with his impure nature cannot attain.

Yet Ulrich's proposal is problematic, for it would break up our moral energy and reduce it to its purely material substratum, 'the infinite complex of the possibilities of life'. But the discoveries of the imagination, from which we seek the desired solution, are themselves forms in which we cannot place our unconditional trust.[36] A more profound suggestion, which borders on the ultimate truth that eliminates the eternal conflict between limit and the unlimited, seems to emerge from Ulrich's discussion with the fanatic Hans Sepp: 'The highest intensity of feeling of which a person is capable . . . is actually a state of rest, of changelessness, like still waters.'[37]

34. Musil, *Der Mann ohne Eigenschaften*, III.11; Italian translation, p. 724. [English version, p. 813.]
35. Ibid., III.12; Italian translation, p. 731. [English version, p. 820.]
36. Ibid., III.38, p. 995. [English version, pp. 1116–17.]
37. Ibid., II.113, p. 542. [English version, p. 608.]

2

Limit

In classical thought, are there any suggestions that the infinite may be realized if it is purged of its negative status as purely potential being?

An initial survey seems to point to the atomists and their attempt to identify a 'minimum': that is, an actual infinitesimal which can resolve the unlimited Becoming of the divisible continuum into Being. We recall Plato's references to our atomistic intuition of space, the atoms of Democritus and Leucippus, and the 'highpoints' (*cacumina*) of Lucretius. But the theories of Democritus and Epicurus were concerned with matter; and the hardness, compactness and solidity of its elementary parts were not meant to represent the essence of what might have been defined as an 'infinitesimal'. Atoms were small, but not infinitely small.

It seems easier, then, to detect an implicit reference to 'infinitesimals' in the attempts of Antiphon and Bryson to square the circle. For example, Antiphon thought he could find the square whose area equalled a given circle, arguing that the smallest arc of its circumference could not be distinguished from the smallest segment of a straight line. One corollary of his proof was the actual existence of an infinity of diverse objects. He argues that it is possible to inscribe in a circle a regular polygon with an arbitrarily large number of sides; and that it is also possible to construct a square with an area equal to that of any regular polygon. Thus, if one indefinitely increases the number of sides of the polygon inscribed in a circle, each of its sides increasingly approximates the arc it subtends; and the area between the polygon and the circumference can be reduced until it assumes an arbitrarily small magnitude.

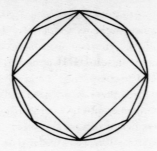

Is it possible at this point to conclude that eventually the polygon will become identical with the circle, and that its sides will be so small as to be considered as arcs, no matter how small, of the circumference? Antiphon maintained that this was clearly possible. The minimal arc of a circumference cannot be distinguished from the minimal segment of a straight line, and therefore a regular polygon with an infinite number of sides cannot be distinguished from a circumference. Moreover, we can construct a square equivalent to a regular polygon with any number of sides, no matter how great. And since a polygon must be considered equal to a circle if its sides are indistinguishable from the arcs of the circumference, Antiphon concluded that it is possible to square the circle.

In reality, arguments of this kind gained little acceptance. In the mainstream of mathematical and metaphysical speculation, thinkers preferred to adhere to the sole meaning granted to the term *apeiron* – its potential meaning. Aristotle backed this view authoritatively when he asserted that Antiphon's reasoning was completely unfounded.

A set of polygons inscribed in a circumference is obviously an unlimited set, since for every polygon with an arbitrarily large number of small sides, there exists a successive polygon with even smaller sides. Yet even this next polygon will not coincide with the circumference, but will admit after itself another polygon. This is a striking example of Aristotle's principle that the infinite is something beyond which there is always something else.

In the same way that the unlimited, or *apeiron*, admits no final term, but only an indefinite development, such a series of polygons cannot reach a conclusive term that coincides with a circumference. If that happened, one would then implicitly admit the real existence of infinite polygons. But this is absurd, for then *apeiron* would be a real infinite, and its intrinsic meaning – traditionally bound to negation, privation and potentiality – would be compromised and unjustifiably altered. In fact, Aristotle maintained, there is no need for mathematicians to introduce actually infinite magnitudes into their proofs, and this does not in the least hinder their prospects for successful research. In the example described above, it suffices to observe that, given any inscribed polygon, it is possible to find another with smaller sides. In other words, it is possible to reduce the area remaining between the polygon and the circle to an arbitrarily small magnitude, without claiming to introduce the problematic concept of an actually infinite set of polygons.

That recognizing the inexhaustibility of the infinite did not compromise the results of mathematical proofs was in fact thoroughly demonstrated by Weierstrass at the end of the nineteenth century. In a rigorous reformulation of the principles of analysis, he proved not only that the infinite could be understood in the potential sense without any loss, but that any reference to this potentiality could be eliminated, or at least concealed, by means of mathematical language that referred only to finite quantities and completely excluded any explicit use of the term 'infinite'. Except for some slightly unorthodox aspects of the heuristic proofs found in Archimedes' *Method*, the Aristotelian point of view was in essence respected in the geometric procedures of Eudoxus, Euclid and even Archimedes himself. Eudoxus' so-called method of exhaustion makes use of the infinite in the only manner envisaged by Aristotle, that is, by avoiding any reference to its presumed actual existence.

Still, the examination of such methods eventually opens the way to a different conception of the infinite.

To do this, we must provisionally think back to Antiphon's arguments. In these, there appears a series of polygons which is not blindly unlimited, but is directed (despite its unlimited nature) towards an end represented by the circumference. Thus, polygons exhibit a recognizable sort of teleological order in which the circumscribed circumference is in some way the unattainable final cause. The circumference is a limit that 'comprises' the unlimited series of polygons although it does not effectively constitute its final term. Still, it offers a solution to the indefinite potentiality of the series to develop, even though it lies *outside* the series.

This means that it is possible to represent concretely the final solution of an unlimited process without denying its potential nature. The inexhaustibility of the unlimited remains an undeniable fact, but it does not force us to accept a mere approximation of what we are trying to attain.

Although the unlimited never departs from the finite, but is merely an interminable repetition of it, what is unlimited can indicate something that transcends it, and can point to this transcendence as a peculiar sign of its actual and infinite completeness, thus evoking or reflecting its essential nature.

For example, let us think of the sum of any converging infinite series. This is a limit which can be rendered concrete in a well-defined mathematical entity, but which does not belong to the infinite series of partial sums that tend towards it. The limit is not the final term of the series, and is therefore not merely an approximation of the result that we are trying to obtain. We attain this limit by abandoning the indefinite analysis of the series that precedes it, and by adopting an external point of reference. (This point of reference remains invisible if we insist on the rigorous verification of its indefinite and unattainable distance.) We are obviously but a step away from the model of ideal representation which is found in every teleological order of the world involving a 'regression to infinity' of its causes – a regression that can only be resolved by postulating a final cause that is actually infinite and cannot be fathomed by our ordinary perception. Aristotle writes that the final cause represents an end which is not conditioned by anything else, but itself conditions the existence of things.[1] If such a final element exists, its solution will be found in the progression to infinity. But by postulating the unconditional infinity of every unlimited series, we challenge the very notion of the Good.

---

1. Aristotle, *Metaphysics* 2.2.994b9.

In the geometry of Eudoxus and Archimedes we find something like the final resolution of an unlimited process. But what we deduce from it is a peculiar and unique intention that is not inherited from the techniques of 'passage to the limit'. These techniques are analogous in a certain sense to the method of exhaustion, and appeared many centuries later in the works of Cauchy and Weierstrass.

Cauchy's terminology still made abundant reference to the infinitesimal, whereas Weierstrass avoided the term but managed to sketch a theory which Hermann Weyl viewed as anticipating the mathematical notion of an actual infinite. By contrast, in Eudoxus' proofs by exhaustion, the meanings of the infinite are concealed, and the term 'infinite' is never mentioned. Beginning with Archimedes' axiom that establishes the magnitudes between which one may define a ratio, and thus a limiting harmony, we proceed by using a *reductio ad absurdum* that disproves any thesis implying the existence of infinitesimal, non-Archimedean magnitudes.[2] For example, to prove the proportionality between two circles C and C' and the squares constructed on their diameters d and d', we advance the absurd hypothesis that the proportion $d^2 : d'^2 = C : X$ holds, where X is different from C'. Next, using the indefinite reducibility of the area between a circle and the successive polygons inscribed in it (seemingly in agreement with Antiphon's methods), we find that X must be equal to C' solely because the difference between X and C' is larger than the area between C' and a polygon with a sufficient number of sides inscribed in it. If it were not possible to find such a polygon, Archimedes' postulate would be false; and we would have to conclude that the proportion in our absurd hypothesis is true only when X differs from C' by an infinitesimal amount. Yet the infinitesimal is never mentioned in the course of the proof. Nor is the proportionality between the circles and the squares explicitly formulated as the limit of an unlimited series of intermediate configurations of polygons.

---

2. According to this postulate, given two unequal magnitudes, where $a < b$, there exists a multiple of the smaller (a) that exceeds the larger (b): thus, $na > b$. But an 'infinitesimal' is the magnitude that does not obey this postulate, for any multiple of it will always be less than any given finite magnitude. Thus, Archimedes' postulate excludes the existence of 'infinitesimal' magnitudes.

Much later, especially after the initial foundation of Analysis, one would try to refer explicitly to infinity as 'something' – be it a differential, a transfinite number, or a point at infinity on a complex plane. Leibniz, Bolzano and Cantor would all attempt this in the belief that mathematics offered the possibility of representing this 'something' as an object marked by indisputable distinctness, and susceptible of being manipulated as a tangible sign by algebraic mechanics. Their attempt succeeded only in part, however, and the opinion remained unshaken that what they sought to represent was destined to remain irreducibly impenetrable.

By contrast, the Greeks preferred to denote this impenetrability as the absurd, making it an inadmissible alternative to their proofs, which sought to discern harmonic ratios and finite proportions. In the proof cited above, if the squares and circles were not proportional, then the infinitesimal would exist. This is the heart of the argument, and an implicit recognition of the negativity of the infinite.

Yet in Archimedes' search for the 'centres of gravity' of geometric figures, there still remained a veiled sense that the unlimited can be finally resolved. His result – which in modern terms is arrived at by the integration of an area or volume using calculus – is reduced to a *point* of equilibrium in a geometric body. For example, if we turn again to Archimedes' calculation of the area of a parabolic segment, it is entirely based on establishing a centre of equilibrium between a triangle and the given segment, when transferred to a convenient part of the plane. When supported, this centre of gravity cancels the imbalances of weight and exerts its power on the infinity of rectilinear segments of which we may imagine the triangle and the parabolic segment are 'composed'.

By means of this equalizing function, a single point effects the balancing of two areas, and ultimately reveals a perfect harmony of ratio between the area of the parabolic segment and the area of the inscribed triangle: the first is ⅓ of the second.

What is important here is a reasoning involving both a point and an infinity of rectilinear segments, and the fact that a geometrical point, which Euclid defined negatively as something without parts, is definitely 'nothing' and 'non-sense'. (Proclus wrote that Euclid's definition reflects Parmenides' intention of describing principles by means of negative propositions. In fact, a principle has a different nature from its effects; and when we negate these effects, we understand better the nature of their transcendent principle.) All the same, the point 'exists' and supports the ordered structure of the visible phenomenon. The reason is that it contains a ratio, a harmonic mediation between two infinities of segments. But any harmonic mediation is merely a limit; and Plotinus would explain that limit is the very thing which causes us to recognize the nature of infinity in any object on which it acts as a delimiting boundary. In his *Enneads* 6.6.3, we read: 'Even if infinity is limited, it is by this very fact infinite [unlimited]; for it is not limit but the unlimited that is bounded; for there is certainly nothing else between limit and unlimited which partakes of the nature of boundary.'[3] In his *Commentary on the First Book of Euclid's Elements* 1.85–96, Proclus describes how the point uniquely combines potentiality and limiting capacity, and how its operant role, as the centre of stellar revolutions and as the pivot of immutable movements, is fused with the impossibility and identity of its substance. The point is a limit, Proclus writes, and yet it possesses the Unlimited in a concealed form, which is why we find it everywhere in geometry. It is the decisive boundary of geometrical figures: for it governs the chain of sequential relations between the Unlimited and what contains it: a body is delimited by a surface, a surface by a line, and finally a line by a point.

As a powerful force of harmony and energy, the point also figures in the magician's sphere. For this reason, Proclus says that the central point of a sphere symbolizes the divine nature of incantation and illusion. This is the source of all formulas of releasing and binding, and every motion of expansion and recession along the radii that extend from the centre to the periphery. We may note in passing that the geometric pole of the cosmic magic wheel mentioned by Proclus (in Greek *iunx*, *iungikos*) evokes the image which Emerson was to employ many centuries later in his essay 'Circles' to describe the circular nature of the life of the soul. For it proceeds from a ring, at first imperceptibly small and infinitesimal, and then expands in depth in successive concen-

---

3. Plotinus, *Enneads* 6.6.3; Italian translation (Bari, 1973), vol. 3, part 1, p. 288.

tric circles which, every time man raises himself up, he uses to delimit and solidify his own temporary status:

> Nature centres into balls,
> And her proud ephemerals,
> Fast to surface and outside,
> Scan the profile of the sphere . . .

Less than a century later, Hugo von Hofmannsthal was to write that 'nothing exists by itself' and that 'everything is completed in circles.' Contrary to Emerson's vision, his novel *Andreas oder die Vereinigten* (1930) recounts an initiatory passage of reunion with the centre, a retrogressive journey through divine incantation that brings him home to the point or 'geometric place' where divergent experiences are fused, to the unity where separations are reunited and maturity is reconciled with infancy. 'True mathematics,' Novalis wrote, 'is the real and proper element of the magician.'[4]

Whenever something unlimited meets a point that limits it, it may change its nature and become a directed tension that postulates an antecedent actuality. In such a case, it is necessary to make distinctions, and not to consider all unlimited things as necessarily equivalent under the aegis of privation (*steresis*) or matter (*hyle*).

There is a classic example in which we must distinguish between different forms of the infinite: Zeno's first argument against motion. According to this argument, anyone who tries to traverse a certain distance will never reach his goal, because he must in fact traverse the infinite series of intervals into which the distance can be halved. Anyone who wants to arrive at 1 starting from 0 must first reach ½, then ¾, then ⅞, and so forth, traversing in series the spatial intervals of distance (a): ½, ¼, ⅛, $\frac{1}{16}$ . . . $\frac{1}{2^n}$ . . . This is clearly impossible, since the intervals are infinite in number.

Yet concrete reality requires that anyone moving at a constant speed must traverse the unit of distance in a finite time. Travelling at a regular speed, without accelerating, the series of intervals (a) is in fact traversed

---

4. Novalis, *Fragments* 1666 (Milan, 1976), p. 416.

in a series of temporal intervals of distances (b): $\frac{1}{2}, \frac{1}{4}, \frac{1}{8}, \frac{1}{16} \ldots \frac{1}{2^n} \ldots$ which is also infinite. But since the partial sum $(\frac{1}{2} + \frac{1}{2^2} + \frac{1}{2^3} + \ldots + \frac{1}{2^n})$ approaches 1, as n approaches infinity, the total time of travel cannot exceed 1.

With the mathematical notion of limit, we may therefore think that we have found a solution to the paradox by resolving the indefiniteness of *apeiron* – which would render the series (a) inexhaustible – in a total and formally complete entity, which is the traversal of the unit of distance within the unit of time. This is also a property characteristic of limit as it was defined in the nineteenth century, especially by Weierstrass. While it retains the notion of process and unlimited potentiality, limit has the power to resolve such potentiality into formal unity. Hence, Hermann Weyl's statement is true: 'At first glance, it might seem as though with the limiting process the rigid Being is definitely resolved into Becoming . . . This appearance is deceptive.'[5]

Nevertheless, more in its *intention* than in its specific dialectical development, Zeno's proof seems impregnable against any refutation that makes use of limit. It introduces a general method of *reductio ad absurdum* that was adopted even by those thinkers who sought to counteract its efficacy. As Borges reminds us, presumably with a touch of irony, Aristotle himself tore apart Zeno's arguments with laconic disdain, only to use them later as a paradigm in refuting Plato. Once we have entered the realm of enigma, it is difficult to avoid the impression that any apparently coherent argument may prove erroneous. 'Truth is a total error,' Novalis wrote, 'just as health is a total disease.'[6] This is an inevitable corollary of the constant correlation between being and appearance, between reality and dream. 'Appearance possesses the fullness of reality,' Simone Weil wrote, 'but as appearance only. As anything other than appearance, it constitutes error.'[7] Hence, we are led to suppose that Zeno chiefly intended *this*, appealing to the disintegrating power of *apeiron* to prove the contradictory nature of motion.

---

5. Hermann Weyl, *Philosophy of Mathematics and Natural Science* (Princeton, 1949), p. 46.
6. Novalis, *Fragments* 203; Italian edition, p. 89.
7. Simone Weil, *Cahiers*, vol. 3 (Paris, 1974), p. 39. [*Notebooks*, trans. Wills, p. 424. Translator's Note.]

As a negative mirror image of the exemplar hidden beyond every representation, *apeiron* could then be, paradoxically, a symbolic reference to God. (By the same token, the negation of an effect offers the best description of the principle that sustains it – which is how Proclus wrote about the geometrical point.) The idea could be extended further, to evil and non-being, even to provide proof that a world without God suggests his distant presence. The first examples to come to mind are the Book of Job, Dostoevsky and Simone Weil. And perhaps the arguments adduced by Nāgārjuna, in the *Madhyamaka kārikā*, to prove the contradictory nature inherent in motion, are closer to Zeno's.

In the final analysis, however, many refutations of Zeno's paradoxes appeal to the principle of limit and to the possibility of actually intuiting in concrete experience (in which the limiting formal principle is always present) what seems incomprehensible only by virtue of a sophistical analysis devised *a posteriori*. Thus Simplicius observed: 'If Zeno were here, we would tell him that the one-in-act is not multiplicity, since in fact, unity belongs to it properly, but multiplicity only potentially. Hence, the same thing is both one and multiple, but in act it possesses only one of these determinations, whereas it never possesses both of them.'[8]

Adolf Grünbaum recently demonstrated that the measured structure of physical time justifies applying the arithmetical theory of limits to the solution of the paradox.[9] Human awareness of time has a base limit of perceptibility, that is, a minimal threshold beyond which temporal intervals vanish into inconceivable smallness. If we consciously tried to contemplate 'all' the intervals of the series (*a*), it would be realized concretely as a countable infinity of mental acts, and the duration of each of these would be larger than the minimal threshold that time allows. But this insuperable 'minimum' is an Archimedean quantity: when added to itself infinite times, it yields an infinite result. Consequently, the mental contemplation of the entire series would result in an impossibly unlimited period of time. This would happen, for

---

8. Simplicius, *Physics* 99.1; Italian translation in *I Presocratici*, 1:290.
9. Adolf Grünbaum, 'Can an infinitude of operations be performed in a finite time?', *British Journal for the Philosophy of Science* 20 (1969): 203–18.

example, if one 'counted' the intervals of (*a*) one by one, assigning to each of them an ordinal number. This would take more time than the necessary minimum just to conceive or pronounce them. (But it is absurd, Aristotle objected, to maintain that whatever moves, moves while counting.[10]) In reality, by raising doubts about the possibility of traversing the interval (0–1), Zeno exploits the unacceptable delay that is implied by reducing the series (*a*) to the corresponding mental acts of the counting process, but he fails to make clear that this process does not reproduce exactly the measurement of the physical time involved in the actual traversal.

Thus, Grünbaum finds Zeno's argument illegitimate because it uses what is basically an inevitable confusion between two incompatible forms of thought. He explains that we do not experience the intervals into which we subdivide the traversal in any measure that corresponds to their actual nature. Rather, we derive our impression of their duration from the time needed for our acts of mental contemplation, which for each fraction of the distance must perforce exceed our minimal threshold or limit. And this impression is wrongfully superimposed on the intervals of time, which are analytically calculated by an act of pure abstraction that completely ignores our observable and tangible experience. Imagining the traversal of an indefinitely small interval thus means encountering a *fiction*, or rather encountering something which transcends the direct representation of the event as governed by the principle of *limit*. And yet it is precisely this *fiction* – something that does not really exist – which turns up in our analysis of what occurs, and is presented as a certain and incontrovertible fact. In reality, the process can be described using the arithmetical strategy of the 'passage to the limit', which avoids the risk of thinking of the intervals (*a*) one by one and of falling into the error of confusing two incompatible lengths of time. But it is also true that this is a case of building a bridge over the void, and one has the unavoidable impression that it is precisely this *nothing* that 'forms' the event. If Zeno's main intention was to reveal this nullity behind the appearance of limit, his arguments would scarcely tally with Grünbaum's.

---

10. Aristotle, *Physics* 8.8.263a–263b.

At any rate, a 'threshold of observability' is a more or less explicit postulate in physics and in the psychology of perception. Quantum mechanics confirms its inevitability. As Max Born, Louis Marcel Brillouin and Karl Popper attest, even in classic mechanics one can define models whose behaviour is undeniably indeterminate. And this circumstance, as Walter Hoering observes, rests on the certain fact that every measurement has a single finite degree of exactitude, and probably an *absolute* lower limit of possible precision.[11] Even our deepest mental instincts need this limit. Erwin Schrödinger has provocatively noted how, even after the principle of superposition was established, people continued to use terminology that was intrinsically linked to the concept of the *stationary states* of energy and to finite exchanges of energy 'quanta'.[12] Couldn't his observation also imply that we have a natural tendency to think in terms of definite representations that last for a minimal time?

Even Aristotle asserted the existence of a minimal interval of time necessary to complete any action – and, we might probably add, to complete any mental operation. He said that we may divide our actual life infinitesimally only at the risk of lapsing into absurdity.[13]

Yet Aristotle's chief objection to Zeno lay in the distinction between infinite-by-addition and infinite-by-division.[14] If we take a unit of length and add it to itself infinite times, we clearly obtain an unlimited distance that cannot be traversed in a finite time. But if we imagine the unlimited in an opposite sense – by dividing the unit of length by halves into infinite intervals – we find that infinity may in a sense be considered exhaustible within a limited time interval.

The infinity of intervals into which the unit of length can be subdivided has a notable property. It is entirely contained within a limited totality that may form the object of an empirical intuition. Hence, matter assigned to the limits of a finite body is visible and tangible in

---

11. Walter Hoering, 'Indeterminism in classical physics', *British Journal for the Philosophy of Science* 20 (1969): 247–55.

12. Erwin Schrödinger, 'Are there quantum jumps?', *British Journal for the Philosophy of Science* 3 (1952): 109–23.

13. Aristotle, *On the Heavens* 2.6.288b33.

14. Aristotle, *Physics* 5.2.233a24.

its entirety, even though it is implicitly impossible to analyse exhaustively all its components.

In his *Critique of Pure Reason* ('Transcendental Dialectic' 2.8–9), Kant made a distinction between the 'progression to infinity' (*progressus in infinitum*), typically exemplified in the infinite-by-division of an empirically intuitable totality, and the 'indefinite progression' (*progressus in indefinitum*), which is subject to no limitations, except for the temporary one assigned to each step before it passes to the next.

If we divide a segment by halves into increasingly small parts, even the smallest parts of the division are actually already present in the whole before the process of division has begun. Their existence is taken for granted, since it is already implied by their belonging to a limited form, from which they cannot escape and in which they will be found when the process of division inevitably reaches them.

We reach the opposite conclusion when we consider another infinity: that of the universe. In this case no empirically intuitable totality exists that contains every object of the infinite series. This series proceeds by the successive addition of objects belonging to an environment that is perennially placed beyond whatever has already been unified by our intuition.

In the first case, Kant writes, the successive elements of the infinite whole are 'found' in a pre-existent totality. In the second case, they are 'sought' outside a partial totality, which is always finite and always attainable. Both infinites are essentially potential, but the *progressus in infinitum* has the advantage of being surrounded by a limit. In this way, it prefigures the nature of every true infinity, which never develops in an unconditional and unlimited way, but can always be referred to a formal and limiting order. The second infinity, that of the magnitude of the universe, was often considered as not being infinite. For, by repeating the finite elements of which it is made up, it can in fact be reckoned as finite rather than infinite. In a series of convincing arguments, Kepler announced decisively that the universe is finite.

Adolf Grünbaum attempted a sort of compromise that was to reduce notably the distance between infinite-by-division and infinite-by-addition. Following Zeno's example, he imagined a purely kinematic model of a machine that could complete an infinite number of operations in a finite time. He constructed this model by hypothesizing a race through the successive intervals ½, ¼, ⅛ . . . which *would be interrupted* at the beginning of each interval. In other words, he described an intermittent race, with an infinite number of stop pauses of increasingly brief duration (¼, ⅛, 1/16 . . . of the unit of time), but such that the total time of the complete run would not exceed that of the 'continuous' run. Despite obvious problems of discontinuity, associated primarily with the accelerations required to achieve the desired level of velocity in the infinite intervals, Grünbaum succeeded in suggesting a certain coherence and possibly in attaining his goal. Naturally, this was at a purely kinematic level, since from the outset it was clear that the energy required by an 'intermittent' runner for such accelerations must turn out to be absurdly infinite.

According to Grünbaum, a race interrupted and resumed an infinite number of times gives us a clear idea of an infinity of '*staccato* operations' (each one independent and occurring successively in time), which are not the result of an *analysis* of a pre-existent unified and uninterrupted run. If we duly exclude *dynamic* complications, this constitutes a counter-example to the absurdity that an infinite aggregate of things can be constructed in finite time. It attempts to reverse the infinite regression (*regressus in infinitum*) into the positive affirmation of *being*, to transform the absurd into the rational, and fiction into reality. The attempt seems possible when viewed from the perspective of Aristotle's assertion, cited above, that the Good exists despite the unattainability of unlimited series, or when viewed from the later perspective of the more explicit arguments of St Thomas or Leibniz. On the other hand, it may be incompatible with what Zeno most likely intended.

None the less, the *regressus in infinitum* of St Thomas's efficient causes – which are perhaps in harmony with Zeno – shows how the supreme act of creation emerges from the absurd, and how the *absurd* effects shifts in perspective that lead to transcendence. In the arguments of Grünbaum, and of others who previously sought to refute Zeno by analogous stratagems, there is no mention of the priority of the absurd or of its demonstrative power. There predominates instead a direct faith in the reality and legitimacy of form, and in its ultimate victory over the unlimited. Hence, some thinkers have even extended the irrepressible metaphysical urgency, which Arthur O. Lovejoy called the 'principle of plenitude', to the threshold of a nearly infinite completeness. And yet even St Thomas introduced the notion of the absurd, only to speedily suppress it in the name of facts and evidence.

In every age, actuality and limits have provided a necessary premise for theories of knowledge, an essential discriminating tool of thought, and an indispensable criterion for conceptual classifications and abstractions. In his *Commentary on the First Book of Euclid's Elements*, Proclus wrote that limiting elements generally preponderate over limited things (over the unlimited) because they are essentially more uniform, more indivisible and even more ancient. And St Thomas repeated this: *'Intellectus noster per prius intellegit indivisibile quam divisibile.'*[15] In this sense, it is fair to say that Aristotle's statement that 'nothing exists that is unlimited' has always remained fundamentally unassailable, as long as *apeiron* is conceived *a priori* as matter and potentiality.

Even in a thinker like Giordano Bruno, who was thought (wrongly) to confuse the indefiniteness of *apeiron* with true infinity, the priority of the actual over the potential was not neglected.[16] Indeed, Bruno wrote that our abstract, logical intellect is free to divide indefinitely a given magnitude, or to form another by means of discourse; but it cannot believe that its creations conform to the nature of things, since the process of the intellect is a 'pure fiction' (*'fingere de nihilo'*).[17] According to Bruno, we should not seek the fundamental reality of the continuum by using the discursive faculty of the abstract imagination, which prefigures any extension as indefinitely divisible. Rather, we must use our *intellect*, or *mens tuens* (viewing mind), which reaches further than any unlimited progression.[18] This indivisible 'minimum' was certainly not the result of an ultimate step attained by imagining an unlimited series. The true infinity represented by it resembled the infinity symbolized by Archimedes' 'centres of gravity', and therefore

---

15. Thomas Aquinas, *Summa theologica*, Part 1, Question 85, Article 8: 'Our intellect understands the indivisible before the divisible.'
16. Cf. Bertrando Spaventa, *Rinascimento, Riforma, Controriforma* (Venice, 1928), p. 229.
17. Ibid., p. 234.
18. Ibid., pp. 234–6.

shared the nature of ratio and hence of limit, which is the true and unique premise and substance of the real. Nothing prevented limit from being represented as a geometric minimum, as a point or as an infinitesimal, as long as these entities could indicate a qualitative essence or a ratio. 'A limit,' Simone Weil perceived, 'is something infinitely small. Limit constitutes the presence, in an order, of the transcendent order, in the form of something infinitely small. Limit is transcendent in relation to what is limited.'[19] It was this very limit that Bruno sought to represent by his geometric minima, in which we find no trace of an unlimited division of the continuum.

In modern times, the exclusive intelligibility of what is actual and therefore regulated by the principle of limit was recognized from different perspectives by philosophers like Descartes, Hume, Whitehead, William James and Wittgenstein.

Descartes observed that the figures that appear in our dreams, like the 'real' objects they resemble, seem to refer to simpler, more general forms and modalities of existence, to which we can hardly deny the attribute of reality and authenticity. Examples include both corporeal nature in general and in its extension, as well as the figures of extended bodies, their size and quantity, and the time and space in which they are immersed.

In these primary points of reference, which Descartes called *res verae* ('true things'), Whitehead perceived the first realization of the timeless and abstract potentiality that precedes the formation of the world – a set of irreducible actual entities beyond which there exists only the sterile 'separation' of absolute potentiality, that is, of nothing.[20]

'When we perceive any attribute,' Descartes had written, 'we therefore conclude that some existing thing or substance, to which it can be attributed, is necessarily present,' for 'every clear and distinct perception is without doubt something, and hence cannot derive its origin from what is nought . . .'[21]

In Descartes' *Meditations*, Whitehead saw a confirmation of the

---

19. Simone Weil, *Cahiers*, vol. 3, p. 112. [*Notebooks*, trans. Wills, p. 484. Translator's Note.]
20. Alfred North Whitehead, *Process and Reality* (New York, 1969), p. 53.
21. Whitehead, *Process and Reality*, p. 54, citing Descartes, *Meditation* 4.

Aristotelian principle which states that nothing exists but what is actual. But what appears actual at a certain moment entails a separation, or 'cutting off' from what on that actual occasion *is not*, because it is not present as a 'datum' in our awareness. And this separation is a sort of exclusive limitation, or simply a limit, to which the complementary opposite is unlimited potentiality. What is given, might not have been; and what is not given, might have been. 'According to the ontological principle,' as Whitehead formulates it, 'everything is positively somewhere in actuality, and in potency everywhere.'[22]

In the recognition of irreducible actual entities, Hume sought a basis and support for the vagueness of sensory impressions, from which he derived all intellectual activity. In Book 1, Part 2 of his *Treatise of Human Nature*, Hume recognized first of all that the capacity of our mind is not unlimited, and therefore that it cannot exhaustively review the indefinite content of a potential infinite. 'The mind cannot attain a full and adequate conception of the infinite.' For encountering it, the mind is forced to follow the interminable successive steps dictated by the inexorable law of becoming, which give us no hope of a definite solution of the process.

For this reason, Hume suggests, the division of ideas must reach an ultimate term, and the imagination must be able to 'reach a "minimum", and may raise up to itself an idea, of which it cannot conceive any sub-division, and which cannot be diminished without a total annihilation'. We may legitimately speak of the thousandth and tenthousandth part of a grain of sand, and succeed in forming a distinct idea of such numbers and their proportions. But the images that the mind succeeds in forming of the initial grain and of the parts into which it is divided are not at all different from each other: their different magnitudes do not alter the quality of the image, which invariably remains that of a grain of sand. The conclusion is obvious: 'The idea of a grain of sand is not distinguishable nor separable into twenty, much less into a thousand, ten thousand, or an infinite number of different ideas.'[23]

---

22. Ibid.
23. Hume, *A Treatise of Human Nature* (Oxford, 1958), p. 27.

Hume further proposes a simple experiment. Let us take a spot of ink and move away from it while keeping our eyes fixed on it. At a certain point, the spot becomes invisible because of the excessive distance; but as it is about to vanish, it will still be visible as a point or indivisible 'minimum'. Just as an ultimate term is given for the ideas of the imagination, he writes, so for sensory impressions we pass from nothing to a minimal actuality which is perceptible and yet irreducible to smaller parts.

The formal and limited character of every experience was also viewed under the aspect of an 'atomization' of reality, in which there comes into play a sort of conscious or unconscious 'decision' that separates what is given from everything that is not given. The primary meaning of 'decision', Whitehead reminds us, is a 'cutting off' or separation. In *The Man Without Qualities*, Musil explained that 'ultimately a thing exists only by virtue of its boundaries, which means by a more or less hostile act against its surroundings . . . There is no getting away from the fact that man's deepest social instinct is his antisocial instinct.'[24]

Atomization initially manifests itself in the perception of the purely spatial or temporal character of all existence. It materializes primordially in temporal rhythms and 'epochs', and in the spatial volumes and forms of objects that move in time.

Like Hume, William James arrived at the extreme threshold of existence as actuality, beyond which he pointed to nothing. 'Either your experience is of no content, of no change,' he wrote, 'or it is of a perceptible amount of content or change. Your acquaintance with reality grows by buds or drops of perception. Intellectually and on reflection you can divide these into components, but as immediately given, they come totally or not at all.'[25]

Wittgenstein's 'simple objects' likewise respond to the demands laid down by the priority of the actual. They are primary entities that can only be decomposed at the cost of 'senselessly' breaking up the 'sense' of the proposition in which they appear. To quote Wittgenstein: 'If the

---

24. Musil, *Der Mann ohne Eigenschaften*, I.7; Italian translation, p. 22. [English version, *The Man Without Qualities*, trans. Wilkins, p. 22. Translator's Note.]
25. Whitehead, *Process and Reality*, p. 84.

complexity of an object is definitive of the sense of the proposition, then it must be portrayed in the proposition to the extent that it determines. And to the extent that its composition is *not* definitive of *this* sense, to this extent the objects of this proposition are *simple*. They *cannot* be further divided . . . since the demand for simple things *is* the demand for definiteness of sense.'[26]

Every proposition thus appears perfectly articulated into simple parts.[27] It is therefore perfectly sensible, even if theoretically it can be decomposed indefinitely, as is implied by that degree of generality and vagueness which nevertheless constitutes its 'form', which is after all its limit. 'For example, if I say that this watch is not in the drawer, there is absolutely no need for it to *follow logically* that a wheel which is in the watch is not in the drawer, for perhaps *I had not the least knowledge* that the wheel was in the watch, and hence could not have meant by "this watch" the complex in which the wheel occurs.'[28] And equally 'if I notice that a spot is round, am I not noticing an infinitely complicated structural property?' But in fact, 'a proposition can quite well treat of infinitely many points without being infinitely complex in a particular sense.'[29] In reality, in every proposition many things are mentally implicit that we do not express, and that may well constitute a convention. But it is indubitable that the sense of a proposition represents the synthetic unity in which is contained and delimited the potentially unlimited whole of the things that are understood.

Thus, a proposition, an object, a living creature or an inorganic mass all bear within their being a sort of perfection that can be linked symbolically to the absolute perfection of the authentic and inscrutable infinite actuality. Each thing is also perfect because it represents a miraculous solution of the potential infinity of the parts into which it can be broken down. In the end, it may be considered a particle of that Good which Aristotle posited as the absolutely final cause of an unlimited process.

Nicholas of Cusa wrote: 'The Infinite Form is only received finitely,

---

26. Wittgenstein, *Notebooks* 18.6.15; Italian translation (Turin, 1968), p. 162.
27. Wittgenstein, *Tractatus* 3.251; Italian translation (Turin, 1968), p. 14.
28. Wittgenstein, *Notebooks* 18.6.15; Italian translation, p. 163.
29. Ibid.

so that every created thing is, as it were, a *finite infinity* or a created god, so that it exists in the way this can best occur, as if the Creator had said: "Let it be made." But since God, who is eternity, could not be made, there was made what could most resemble God. It follows from this that every created thing *qua* created thing is perfect, even if it seems less perfect than another.'[30]

All the same, this Renaissance affirmation of formal completeness probably conceals a more subtle relation between being and non-being, which is instead revealed in Nicholas of Cusa's mathematical works and in his other statements where the descriptive art of paradox predominates. These passages give the impression of a possible final victory of the potential over the actual, of absence over formal presence. If we sought decisive confirmation of this priority, we would find it in certain statements of the *Tao te Ching*. 'We join thirty spokes at the hub to make a wheel; the usefulness of the carriage depends on what is not there.' 'The usefulness of the vessel depends on what is not there.' 'We cut doors and windows to make a house; the usefulness of the house depends on what is not there.' These sudden flashes of cognitive intuition – both elementary and unexpected, and subverting our common sense – foreshadow the description of an ethical miracle, namely, the condition of the saint. 'He who knows white but cleaves to black is the measure of the world.' 'He who knows he is a tomcat but behaves like a hen becomes like a ravine. The power [tê] never leaves him. He returns to the state of the suckling child.'[31] By a disquieting association of ideas, this calls to mind an exclamation by Hans Sepp, who embodies the dangerous madness that arose in the declining Austria of *The Man Without Qualities*. The purest moral ideal for Hans Sepp is the condition of a child at the breast, which is marked by the greatest potentiality and the highest concentration of spontaneity and creative energy.

The same risk also appears in the prefigurations of death given a certain prominence by D'Annunzio. But even here we must not over-

---

30. Nicholas of Cusa, *De docta ignorantia* 2.2; Italian translation, *La dotta ignoranza* in *Opere filosofiche* (Turin, 1972), p. 114. [English version, *On Learned Ignorance*, trans. Jasper Hopkins (Minneapolis, 1981), p. 93. Translator's Note.]
31. *Tao te Ching* 11 and 28; Italian translation, pp. 49, 80.

look several moments of happy and reassuring intuition, as when Stelio Effrena in *Il Fuoco* speaks of music: 'Have you ever thought that the essence of music is not in its sounds? . . . It is in the silence that precedes the sounds and in the silence that follows them. Rhythm appears and lives in these intervals of silence.' Simone Weil, inspired by Valéry, makes a similar observation: *'La musique part du silence et y retourne . . . création et durée.'*[32] And another synthesis of this truth is summarized in a fragment of Novalis: 'Everything visible is connected to the invisible, the audible to the non-audible, the sensible to the non-sensible. Perhaps the thinkable to the unthinkable.'[33]

---

32. Simone Weil, *Cahiers*, vol. 1 (Paris, 1970), p. 94. [*Notebooks*, trans. Wills, p. 2: 'Music starts from silence and goes back to it . . . creation and duration.' Translator's Note.]

33. Novalis, *Fragments* 1710; Italian translation, p. 428.

3

Irrational Numbers

The truth which states that nothing exists except what is actual would seem to be contradicted by the fact of the existence of irrational numbers. But what existence does this refer to? An irrational number – that is, a number that cannot be represented as the ratio between two integers – may never be written as a finite sequence of digits. If we try to express it explicitly, it displays digits that, after a certain point, observe no apparent order or formal law. The digits seem to be randomly distributed, as would be the case with successive indefinite rolls of the dice.[1] In its first manifestation, the irrational is configured as a pure impossibility: one cannot, for example, represent as a rational fraction the ratio between the side and the diagonal of any square. This impossibility is designated by the symbol $\sqrt{2}$. (Indeed, $\sqrt{2}$ is the non-existent numerical ratio between the lengths of the diagonal and the side of a square, with reference to a common unit of measure. Hence, such numbers are called 'incommensurable'.) Another impossibility, that of defining a rational fraction between the lengths of a circle's circumference and its radius, is represented by the symbol π.

We would not be far from the truth if we maintained that an irrational number does not in fact exist, even if we retain a deep-seated certainty that it undoubtedly designates 'something'. An irrational number seems to have no actual existence; it cannot be expressed as the set of 'all' its digits.

Aristotle attributed the discovery of incommensurable geometrical

---

1. Yet the digits of the number π form a 'curious repetition pattern': cf. Donald Knuth, *The Art of Computer Programming*, vol. 2 (Reading, Mass., 1969), p. 35.

magnitudes to the Pythagoreans, and the note of mystery and inscrutability implicit in the discovery is adumbrated in Plato's dialogues. In
point of fact, we know very little about this 'discovery', and it is likely
that incommensurable magnitudes were already known two or three
centuries before Pythagoras. The problem is made even more complex
by their presence in the texts of the *Śulvasūtra* (written between
the seventh and second centuries BC), which describe geometrical
procedures for building the altars used in Vedic religious rites.[2]

In order to understand the sense attributed to the irrational, we must
in fact extend our investigation beyond the traditionally respected
boundaries. In Plato's reflections on the essence of number, we may
detect the remnants of an archaic vision of reality which could likely
be interpreted in terms of temporal rhythms and their numerical ratios.
What we would discover is described in a passage of *Hamlet's Mill* by
Giorgio de Santillana and Hertha von Dechend, where we read that
'the order of Number and Time was a total order, to which all belonged,
gods and men and animals, trees and crystals and even absurd errant
stars, all subject to law and measure.'[3] The authors clearly show that
a cosmic order based on number was not the prerogative of the Greeks
alone. Hence, in retracing this order, it is useful to re-examine the
mythical symbology of the poems of Snorri Sturluson and of Scaldic
poetry, of the ancient epic poems of Finland and Estonia, of the Iranian
or Avestan national epic, of the Hindu *Mahābhārata*, and of the
mythologies of the American Indians.

The connections revealed by De Santillana between archaic myth
and the numerical proportions of astronomical cycles can also be
discerned in Plato's dialogues, where we find ourselves on the threshold
of an explicit philosophical language. In his *Epinomis* (978B), Plato
sees the origin of counting in the fundamental laws of celestial phenomena. Uranus, the god who gave us the wisdom of number, makes clear
the primal distinction between one and two by means of the indefinite
alternation of day and night, while the calculation of precise numerical

---

2. See C. N. Srinivasiengar, *The History of Ancient Indian Mathematics* (Calcutta,
1967). See also Abraham Seidenberg, 'The Ritual Origin of Geometry', *Archive for
History of Exact Sciences* 1 (1962): 488–527.

3. Giorgio de Santillana and Hertha von Dechend, *Hamlet's Mill* (Boston, 1969), p. 6.

ratios arithmetically reproduces the more complicated ratios of solar and lunar months.

In turn, the descriptive power of numerical proportions is extended from their original model in astronomical occurrences to all natural phenomena. In his *Philebus*, Plato explains that the optimal state of a form and the stable character of an object are expressed in a well-defined proportion that fixes a limit-point at which an object's indefinite capacity to evolve is resolved and beyond which it is useless to proceed. For example, the healthy condition of a human being is the result of an optimal relation between wet and dry. Once this is realized, there is no sense in exceeding this limit by indefinitely continuing whatever medical treatment has been administered.

Every development is a 'birth into being' (*genesis eis ousian*), a straining towards some type of stability, when the 'more or less' or the 'large or small' (*mega kai mikron*), not yet properly regulated, alternate in infinite oscillations before reaching equilibrium. The 'more or less' is simply Plato's version of Aristotle's *apeiron*, on which limit (*peras*) must act in order to generate that 'mixture' through which all existence is realized.[4]

Even the science of discourse is based on number, or *logos*, a word which also means analogy, relation, correspondence, and is synonymous with ratio, specifically with numerical ratio. If we ignore number, Plato teaches us, we lose the very faculty of thinking and judging, and live only by our sensations and recollections. Similarly, Nicholas of Cusa comments in his *De docta ignorantia* (*On Learned Ignorance*) 1.1: 'All those who carry out an investigation judge the uncertain proportionately, by means of a comparison with what is certain. Therefore, every inquiry is comparative, and uses the means of proportion . . . Therefore, number, which is a necessary condition of proportion, is present not only in quantity, but also in all things which can in any way be similar or different, whether substantially or accidentally.' Hence it follows that 'the infinite, because it is infinite and eludes every proportion, is unknown.'[5]

---

4. See A. E. Taylor, *Plato: The Man and his Work* (Cleveland, 1966), pp. 414–15.
5. Nicholas of Cusa, *De docta ignorantia* 1.1; Italian translation, p. 57; [English translation, p. 50].

But irrational numbers likewise defy every notion of ratio and hence of proportion; and since the irrational is accordingly unknown and unknowable, we may ask if it can legitimately be counted as a number. An affirmative answer to this question was given by Richard Dedekind in 1872, based on the theory of proportions previously advanced by Eudoxus and Euclid. But even Plato probably had a similar conception, and Dedekind's procedure is perfectly compatible with the Platonic conception of number as a synthesis of the One and the Dyad. Indeed, Plato had asserted that numbers are the forms that result from two constituents: the One, or limiting principle, and the indefinite Dyad (*aoristos dyas*), which is identical to *apeiron*, or the principle of more or less. Evidently, what unmistakably determines a number as such is its formal constituent, that is, the One which is present at every step of the indefinite process of counting.

If we now construct the series (1) of successive fractions

$$(1) \qquad 1, 1 + \frac{1}{2}, 1 + \frac{1}{2 + \frac{1}{2}}, 1 + \frac{1}{2 + \frac{1}{2 + \frac{1}{2}}}, 1 + \frac{1}{2 + \frac{1}{2 + \frac{1}{2 + \frac{1}{2}}}}, \ldots$$

we obtain values alternately more or less than $\sqrt{2}$, and each successive fraction moves closer to $\sqrt{2}$ than the preceding one. As it is extended to infinity, the series (1) converges towards $\sqrt{2}$, in the sense that, given any quantity ε, all its elements, except for a finite number, raised to the square differ from 2 by a quantity less than ε. Thus, what we obtain is an indefinite approach to determining (through the principle of One) the indefinite Dyad represented by the infinite oscillations of the series (1) around $\sqrt{2}$. Hence, even the irrational is, at its limit, a number, and it was as such perhaps that Plato imagined it, unlike Aristotle and Eudoxus, who usually represented incommensurables only by means of geometric magnitudes.[6] A. E. Taylor showed that the procedure of approximating $\sqrt{2}$ by means of the series (1) was in all likelihood known to Plato, who in his *Republic* (546c) refers to the numerators and denominators of its

6. See Taylor, *Plato*, p. 505.

successive fractions, which came to be called *pleurikoi kai diametrikoi arithmoi*. Both Theon of Smyrna and Proclus also described analogous findings.

None of this contradicts in the least the following interpretive model provided by Simone Weil in her *Notebooks*.[7] Let us imagine that we place two groups of weights on opposite sides of a balance, and that the weights are formed of cubes, with the sides of the first group equal to the diagonals of the second. No number of weights on either side could achieve equilibrium. And all the possible ratios between the number of weights placed in the first and second groups would then be distributed in two classes, depending on whether the balance tipped to one side or the other. All these ratios would represent a progressive approximation towards an ideal point of mediation, but one which is not based on any real or concrete existence in the experiment.

This non-existent equilibrium, Simone Weil points out, is a *logos alogos* (irrational ratio), a reality in some way existent but unfathomable and defying any effective denomination. Conceptually, it parallels our representation of all truth as an unnameable point, or Greek *alogos*, around which we may position all possible views of a topic by locating them in their proper place.[8] But the point of mediation *par excellence*, the ultimate truth which eliminates all the imbalances and oscillations that reflect the eternal polarity of limit and unlimited, is revealed in a fragment of Anaximander: 'The principle of beings is the indefinite . . . From that place where beings have their origin, they also of necessity have their passing away: for they pay one another the penalty that expiates injustice according to the order of time.' Thus, the impenetrable justice and general equilibrium that govern the perennial alternation, in material reality, of reciprocal imbalances coincide with the absolute infinity or *apeiron* of Anaximander.

We are indebted to Richard Dedekind, and to parallel attempts by Cauchy, Weierstrass and Cantor, for a rigorously mathematical proof

---

7. Simone Weil, *Cahiers*, vol. 2 (Paris, 1972), pp. 31–2. [*Notebooks*, trans. Wills, p. 162. Translator's Note.]

8. Ibid., p. 32. [English translation, p. 162.]

showing that it is possible to define the irrational using an unambiguous criterion of determination – that is, in Platonic terms, using a determination of the indefinite Dyad based on the One.

In a first outline of his findings, Dedekind began by recognizing the analogy between rational numbers and the points of a straight line. This analogy changes into a precise correspondence if we mark a point o and one unit of measure on a line, and if we then create a point P on the line, which corresponds to a rational number a, and whose distance from o is traversed by the segment determined by a. Yet by posing the question in these terms, we immediately find that for infinite points on the line there is no corresponding rational number: it is enough to consider the case of point Q in the figure below.

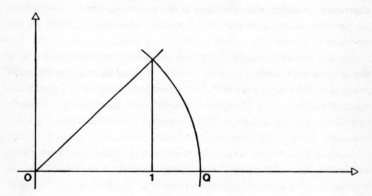

In other words, the straight line's characteristics of 'continuity' and 'completeness' – to use these terms in the sense momentarily suggested by intuition – are not reproduced by the domain of rational numbers, in which there are 'holes' (a horribly ambiguous expression) which render problematic any attempt to express the geometric continuum arithmetically by using rational numbers alone.

So, in order to reach a definition of 'real' number that includes what is 'missing' from the domain of rational numbers, Dedekind decided to take into consideration two problems:

1. To establish axiomatically a property that characterizes the continuity of a straight line.

2. To define 'real' numbers so that they conform to an arithmetical version of the continuity of a straight line.

The first problem translated into the following principle. If all the points of the straight line are comprised in two classes such that every point of the first class lies to the left of every point of the second class, only one point exists that determines this division of all points into two classes, and thus divides the line in two parts. Each point of the line is uniquely determined by the section produced by it and therefore can be appropriately called a 'cut' (German *Schnitt*, 'section').

The second problem was solved by observing that every rational number x designates by analogy a 'cut' in the ordinary rational domain, that is, a pair of classes A_1 and A_2 such that every number of A_1 is less than every number of A_2, while x is the largest number of class A_1 or the smallest number of class A_2. Yet even when defined without any ambiguity, a pair (A_1, A_2) such that every number of A_1 is less than every number of A_2 may not correspond to any rational number x, in the same way in which a cut in the line (such as the one produced by point Q in the figure) may not correspond to any rational fraction. But whenever this occurs, Dedekind thought, it is legitimate to 'create' a new (irrational) number y which corresponds to the pair (A_1, A_2), meaning that it *is* itself the pair (A_1, A_2).

A. E. Taylor observed that this 'cut' is in perfect harmony with the Platonic notion of number.[9] For it is a determination, effected by the principle of the One, of the 'more and less' implicit in the classes A_1 and A_2. It is an operation produced and legitimized by the fact that the pair (A_1, A_2) can be uniquely defined.

The property of continuity in the real domain was thus linked to the continuity of the straight line, even though the arithmetic theory of real numbers was completely independent of any possible geometrical illustration.

In this way, an irrational number seems even more 'real', insofar as it can be referred to a point on the line. For example, point Q seems to corroborate the hypothesis that the number $\sqrt{2}$ actually exists, even if defining $\sqrt{2}$ does not depend at all on its visualization as a geometric

---

9. Taylor, *Plato*, p. 511.

point. Yet if we try to conceive of an irrational number as an actual, concrete entity, we encounter all kinds of paradoxes. The irrational is not an actual infinite in a categorical sense, but rather the analogue of anything posited as the invisible resolution of a process that is both unlimited and teleologically ordered. It is not a true infinite, which is mathematically non-representable, but an analogue of that transfinity which was later described by Cantor as a notion that merely hints at absolute infinity.

Besides, the geometric point seems even less definable than an irrational number; and our sense that it 'belongs' to a straight line is even less comprehensible. We need merely think of what would happen if we 'removed' point Q from the line on which it is located. Would we really create a 'hole' between two points, through which the circumference that formerly intersected the line at Q would now pass? In reality, as Wittgenstein wrote, 'the expression "straight line with a point missing" is a fearfully misleading picture.'[10] For 'the picture of the number line is an absolutely natural one only up to a certain point; that is to say so long as it is not used for a general theory of real numbers.'[11]

Hence, it is convenient to regard Dedekind's section less as a 'cut' indicating an effective localization of number, than as a successive approximation to the two adjacent limits of the lower class A_1 and the upper class A_2.[12] As a matter of principle, one may question whether it is possible to represent the limit of this process as a verifiable entity, such as a geometric point would appear to be. The irrational number is the process itself, by virtue of the subtle fact that the classes A_1 and A_2 can be differentiated – through that process – in an unambiguous way, and are in fact different when they correspond to different numbers.

---

10. Wittgenstein, *Bemerkungen über die Grundlagen der Mathematik*; Italian translation, *Osservazioni sui fondamenti della matematica* (Turin, 1971), p. 198. [English version, *Remarks on the Foundations of Mathematics*, trans. G. E. M. Anscombe (Cambridge, MA, 1978), p. 291. Translator's Note.]

11. Wittgenstein, *Osservazioni*, p. 195. [*Remarks*, p. 286. Translator's Note.]

12. Wittgenstein, *Osservazioni*, p. 197. [*Remarks*, p. 289: 'In Dedekind we do not make a cut by cutting, i.e. pointing to the place, but . . . by approaching the adjacent ends of the upper and lower class.' Translator's Note.]

Dedekind himself pointed out that the inexorable tendency to consider space solely as a continuum is in part unjustified. All the geometrical constructions that appear in Euclid's *Elements*, he noted, could be adapted to a purely rational space, whose discontinuity would in fact pass unobserved.

If we try to define the continuum as the result of simple, indivisible and actual parts, it is transformed into a model of pure indefiniteness, into a receptacle for all the *aporiai*, or perplexities, that *apeiron* engenders as soon as we claim to display it in concrete forms. Whitehead saw the ideal model of pure potentiality in the continuum, and he regarded atomism in turn as the prototype of actuality.[13] The attempt to base the former on the latter – that is, the continuum on the presumed actuality of punctual components – could only be made by explicitly invoking transcendence, or at least a free creation of entities which cannot be reduced to pure rationality or to the ordinary mathematical apparatus deducible from whole numbers. For no finite combination of symbols *per se* can constitute a representation of the infinite; such a combination is always preceded by an idea of what it seeks to designate.

The attribute of 'negativity', traditionally assigned to points in geometric space, was in harmony with this transcendence, and was at the same time an indispensable prerequisite for any paradoxically symbolic reference to the absolute. In fact, as happens in viewing a cube – we can never see all its sides at the same time, but we still have an idea of its unity – what seemingly lies hidden may prove to be the principal reason for its ordered structure. The explicit designation of this negativity as a visible or rationally definable object, or the transformation of *apeiron* into an articulated name or indeed into an algebraic sign, often collapsed into paradox or into mere inconclusiveness. The mathematical progress derived from it is clearly a different matter, and should not be considered a real approximation to the idea of the absolute. As a result, there is some value in highlighting the negative aspects of the continuum by reviewing some well-known pronouncements on the subject.

Early on, Plato established that points are a 'fiction' of geometricians, and he assigned them the sole property of delimiting the segments of a

---

13. Whitehead, *Process and Reality*, p. 76.

straight line. Subsequent Platonists, to express the impossibility of reducing the continuum to an actual set of indivisibles, spoke of the line as the 'flux' (*rhysis*) of a point, an expression which Newton later used in his theory of the infinitesimal.[14]

Arguing against the atomists, Aristotle denied that continuity arises from the juxtaposition of contiguous points. He formulated a view later described mathematically in the intuitionism of Brouwer in the first decades of the twentieth century. The continuum cannot be described by means of a relation that binds a set to its elements (the continuum has no simple constituent parts), but by means of the concepts of 'part' and 'whole': a portion of a segment is a 'part' of the segment conceived as a 'whole', rather than one of its constituent elements.

Roger Bacon observed in his *Opus maius* that, if lines were formed of points, the ratio between the side and the diagonal of a square would prove equivalent to a ratio between integers and thus, absurdly, to a rational number.[15]

Duns Scotus added a further proof. A given number of concentric circles will all be intersected by any radius drawn from their common centre. Now, if the circles were formed of contiguous points or some other kind of indivisible elements, these would be equal in number for every circle, which is manifestly absurd.[16]

St Thomas Aquinas described the limit-like nature of each point on a straight line as follows: '*Manifestum est enim quod in linea infinita, et etiam in circulari, non est punctum nisi in potentia.*'[17] By verifying that the intellect conceives *a priori* only what is actual, St Thomas reaches this conclusion: '*punctum non ponitur in definitione lineae communiter sumptae.*'[18]

William of Ockham substantially agreed with Aristotle. He distin-

---

14. Taylor, *Plato*, pp. 505–6.
15. Pierre Duhem, *Le système du monde. Histoire des doctrines cosmologiques de Platon à Copernic*, 10 vols. (Paris, 1954–73), vol. 7, p. 21.
16. Ibid.
17. Thomas Aquinas, *Summa theologica*, Part 1, Question 85, Article 8. ['It is manifest that in an infinite or circular line, a point only exists potentially.' Translator's Note.]
18. Ibid. ['A point is not included in the definition of a line as commonly assumed.' Translator's Note.]

guished between 'continuity' and 'contiguity', and denied that the former could be broken down into a juxtaposition of contiguous indivisible parts, meaning lines. The same applies to the continuity of time, which cannot be broken down into a composition of instants: 'An instant is not an absolute entity such that it can be distinguished from every divisible entity.'[19] Just as the actuality of a continuum does not consist in the parts into which it is divisible, so 'time exists in the entire movement of the entire heaven, and not merely in some part of movement in some part of heaven.'[20]

The philosophers of the thirteenth and fourteenth centuries substantially agreed with these conclusions. Most of them said it was impossible to break down the continuum into a set of points. Some, including William of Ockham, even denied that the point functioned as the boundary delimiting finite segments of a line: it is not necessary to grant the existence of a point because a magnitude has a termination.

Gregory of Rimini called points, lines and surfaces 'fictitious magnitudes', and observed that the geometrician does not need to suppose that they really exist: they are creations of the spirit.[21]

Buridan observed that, when we assert that a point is indivisible, we don't mean that it is 'really' so, nor that we should take this literally.[22] When astronomy advances the hypothesis of epicycles and eccentrics, it employs fictions useful in calculating the position of the planets, and nothing more. In any case, by using points, lines and surfaces, we may construct a fictitious geometry whose exemplary clarity aids our intellect in better understanding reality.

Thus, nominalist trends of thought anticipated modern idealism, and the image they provided of the continuum was not very different from that described by Kant's formula: the law of the continuum of forms is a simple idea, to which a congruous object of our experience cannot be referred.

---

19. William of Ockham, *De sacramento altaris*, ed. T. B. Birch, cited in T. B. Birch, 'The theory of continuity of William of Ockham', *Philosophy of Science* 3 (1936): 494–505.
20. Ibid.
21. Duhem, *Le système du monde*, vol. 7, p. 33.
22. Ibid., pp. 37–8.

Pierre Duhem showed how Buridan, Walter Burley and Albert of Saxony developed ideas not unlike those that led Dedekind to conceive a definition of real number as the separator between two unlimited series.[23] The problem was a variant of those discussed so far. The point of equilibrium between a force and an opposing resistance is defined by two kinds of outcome: one predicts motion in the direction of the force; the other that the force is overcome by the resistance. Both series of outcomes can be brought indefinitely close to the point of equilibrium, by running through the continuum of the infinite intermediate possibilities. The point of equilibrium is a limit point that belongs to neither of the two infinite series which define it.

Leibniz in turn conceived the mathematical continuum as an ideal entity, that is, as an imaginary one, and developed a relational theory of space, assigning to geometric points a function analogous to that conceived earlier by Plato.

At the end of the nineteenth century, Cantor offered a mathematically reliable definition of the continuum, which Russell viewed as refuting the traditional philosophical view that ruled out regarding the continuum as a set of points. Still, Cantor proved that the continuum is not countable, that is, that not all of its constituent elements can be counted, one by one, using whole numbers. Writing in 1909, Émile Borel interpreted Cantor's conclusion as proving the existence of an undefined component of the continuum. 'In the geometric continuum,' he writes, 'there clearly exist elements that cannot be defined: this is the real meaning of Georg Cantor's important and celebrated assertion that the continuum is not countable. The day on which such undefinable elements are actually put aside and are no longer more or less implicitly used will see a vast simplification in the methods of Analysis.'[24] Elsewhere, Borel remarked that the technique by which Weierstrass had reformulated Analysis is based on the commonly accepted fact that all forms of mathematics can be deduced from the unique notion of whole number: 'The fundamental notions involving the idea

---

23. Ibid., p. 83.
24. Émile Borel, 'Les probabilités dénombrables et leurs applications arithmétiques', *Rendiconti del Circolo Matematico di Palermo* 27 (1909): 247–71.

of limit (incommensurable numbers, derivatives, definite integrals, integrals of differential equations, etc.) are defined from whole number: once we have acquired such notions, it is possible to use them without necessarily involving in each case their definition by means of whole numbers.'[25]

While mathematically unimpeachable, Cantor's definition did not exhaust the aggregate of concepts and images that an intuition of the continuum had always evoked. In particular, the idea of unlimited potentiality, which Cantor's vision had certainly not privileged, re-appeared in Brouwer's intuitionistic mathematics. In this new theory, the continuum again began to resemble what Aristotelian and Thomist philosophy had already envisioned.

In the localized points of the continuum (and in particular in irrational numbers), Brouwer perceived not so much actually definable entities as various series of natural numbers succeeding each other in the order decreed by an arithmetical law or by a 'free choice'. Geometrically, the idea could then be represented as the determination of a point by means of an infinite series of intervals characterized by a 'level' of continually increasing smallness. The coincidence of two points signified the coincidence, at a certain point, of the two series that represented them. By contrast, the distinction between them was revealed by the discovery, in the two series, of two separate intervals, one containing the first point and the other the second.

But recognizing this equality or distinction could not be taken for granted. The exclusion by the intuitionistic mathematics of the principle of the 'excluded third' determined the impossibility of a solution in principle that was independent of the effective mathematical construction, of the problem of the equality or inequality of the points. To say *a priori* that two irrational numbers (or two points) must be either equal or distinct was nonsense, so that one had to recognize a third logical possibility – between the 'yes and no' of the principle *tertium non datur*. The dilemma thus codified in the rejection of the 'excluded third' – of which scientific literature offers

---

25. Émile Borel, 'Contribution à l'analyse arithmétique du continu', *Journal de mathématiques pures et appliquées*, 5th series, vol. 9 (1903): 329–75.

numerous examples referring to the case of the most familiar irrational numbers – perfectly described the intuitive characteristic of the continuum as a site of the indefinite. In particular, the idea of 'contact' between two points (a notion whose problematic nature was often emphasized by Aristotle) was now recognized in its aspect of legitimate ambiguity.[26]

It was also recognized by some that the indefiniteness of the spatial and temporal continuum blurs the clarity of outline of the objects that are embedded in it. Husserl explained that the apprehension (idea) of every 'real' thing emerges only as a limit point in the continuum of the process of learning and of our ever more exact experiences of that thing.[27] Phenomenal experience implies in itself the infinity and potentiality of a 'continuum' of appearances; but it is, at the same time, true that the objectivity of ideas (like that of points and irrational numbers) is not consequently compromised: 'The idea of an infinity . . . is not itself an infinity.'[28] 'The immanent being is given to our awareness only *qua* idea,' Husserl further writes, 'since it requires a process of approximation. Adequate givenness is an idea that has the character of a limit, which one may approximate at will.'[29]

Thus, Husserl too seems to allude to Nicholas of Cusa's 'finite infinity', which cannot be described analytically but whose concept can be represented synthetically. Still, this in no way implies that a set can become effectively and actually infinite in an immanent, analytical survey of all the elements found in it.

All the same, it would not be advisable to consider solely the pure negativity of the continuum. Our perspective in contemplating the infinite can always be converted into its opposite. If Non-being seems to be the principle of disintegration in the world of limited forms, it can also be reversed to its apparent opposite, pure Being. By resorting

---

26. Hermann Weyl, *Philosophy of Mathematics*, pp. 52–4.

27. Ibid., p. 41.

28. Edmund Husserl, *Ideen zu einer reinen Phänamenologie und phänomenologischen Philosophie*, sec. 143; Italian translation, *Idee per una fenomenologia pura e per una filosofia fenomenologica* (Turin, 1965), p. 320. [English version, *Ideas*, trans. W. R. Boyce Gibson (London, 1931), p. 398. Translator's Note.]

29. Ibid., p. 388.

to Hegel's proof of the identity of being and nothingness, we may explain similar reversals of viewpoint.

In the nineteenth century, Vincenzo Gioberti taught us, from a theological perspective, to perceive in the continuum the clearest image of what he called the state of *methexis*. This Greek word refers to the 'participation' in divine perfection and to the link between such perfection and the finite creature: thus, it is a synonym of the creative act of God, of the passage from Being to the existent effected in the active Word as an interval between points in space and instants in time. The urgent need for compatibility between the actual infinite and the created universe became a corollary of the mysteries of the creation and incarnation. What Dedekind would call a 'cut' in rational numbers, the *logos alogos* (irrational ratio) of the ancients, was further transformed into the archetype of the *relation* or unique form of accessibility to God's perfection. 'This relation is the bond between things,' Gioberti wrote, 'the unity of the universe, the intelligibility of existences, light, etc. It is the mediator between God and man, between the world and God. It becomes one with the creative act. Christ is the relation personified . . .'[30] Hence, 'the creative act is a passage: from Being to the existent. This passage is infinite; it is an actual infinite; it is the infinite of divine omnipotence. This omnipotence manifests itself finitely in its effect, considered *per se*, as actually finite; it manifests itself in an infinite way, through the mode that produces it, through the infinite interval that it surpasses in producing it.'[31]

If we represent the creative act as the point that becomes a line as it moves, or as the centre that irradiates the infinitely swift movement of the first angelic circle, as Dante described it in Canto 28 of his *Paradiso*, then the decisive proof of God's existence, according to Gioberti, is provided by the philosophy of infinitesimals. The creative impulse is realized in the continuum where, more than anywhere else, we may recognize the invisibility of contact, of relation, of the dialectic bond of opposites. Gioberti sketches here a theological justification of Leibniz's

---

30. Vincenzo Gioberti, *Della Protologia*, vol. 1 (Naples, 1864), p. 177.
31. Ibid., p. 263.

calculus as confirmation that 'mathematical creation is the most vivid image of divine creation'.[32]

To quote Gioberti again: 'The mean between two or more things, their juncture, union, transit, passage, crossing, interval, distance, bond and contact – all these are mysterious, for they are rooted in the continuum, in the infinite. The interval that runs between one idea and another, one thing and another, is infinite, and can only be surpassed by the creative act. This is why the dynamic moment and dialectic concept of the mean are no less mysterious than those of the beginning and the end. The mean is a union of two diverse and opposite things in a unity. It is an essentially dialectic concept, and involves an apparent contradiction, namely, the identity of the one and the many, of the same and the diverse. This unity is simple and composite; it is unity and synthesis and harmony. It shares in two extremes without being one or the other. It is the continuum, and therefore infinite. Now, the infinite identically uniting contraries clarifies the nature of the interval. In motion, in time, in space, in concepts, the discrete is easy to grasp, because it is finite. The continuum and the interval are mysterious, because they are infinite.'[33]

If Gioberti had known the position of intuitionistic mathematics on the principle of the 'excluded third' (or excluded middle), he might have perceived in the third logical possibility between 'yes and no' the mystery of the mediation between every polar opposition. In this context, Dante's verse 'For no and yes debate within my head' (*Inferno* 8.111), which he viewed as symbolizing the dynamic evolution between the one and the many, could well be transferred to every approximation of the irrational that evolves through the indefinite alternation between large and small (*mega kai mikron*). The idea of Lysis the Pythagorean – that God could be conceived as an irrational number[34] – would then assume a meaning perfectly analogous to that suggested by a third logical possibility for the problem of determining an irrational number *a priori*, as the intuitionists would later conceive it. Where the

---

32. Ibid., p. 292.
33. Ibid., p. 160.
34. See Sabètai Unguru, 'Incommensurability and Irrationality: A New Historical Interpretation', *History of Science* 15 (1977): 217.

intuitionists' notion inclines towards the impossibility of an actual infinite, Gioberti would gladly have seen its nearly tangible reflection.

The actual infinite is thus reflected symbolically in everything that plays a mediating role. The verb is transferred by analogy into both language and the spoken word (*langue* and *parole*); and nouns reproduce its ordering power. 'The word,' Gioberti writes, 'is thus the circumscription of an infinite thought. It is therefore in the orders of thought what the creation was in the Pythagorean and Platonic sense of the orders of the Universe; for intuitive knowledge is almost (*quasi*) the matter, or *apeiron* of the intelligible world.'[35]

In the immanent generative act as well, we may discern a resemblance to the archetype: in Latin *penitus* and *penes* signify proximity and intimacy, and also denote generation. From the same root in *peninsula*, *penumbra* and *penultimate*, and in the notion of similarity or 'almost' (*quasi*) suggested by these words, Gioberti perceives the imprint of invisible unity and exemplary idea, of a coordinating identity, and of a dialectic centre that reconciles and harmonizes.

In the visible and invisible link between things, in the connective tissue of the real, we recognize the imprint of the actual infinite: the near-thing or similarity 'on the one hand borders on identity, and on the other introduces an infinite diversity, because the minimum is maximum, both of them being infinite. The reason for this apparent contradiction is the infinity of the idea, of which created things are copies. Being one and infinite, the idea is infinitely similar to and dissimilar from itself, and this infinite similarity and dissimilarity is the archetype of the finite similarity and dissimilarity of creatures among themselves.'[36]

The dynamic infinite that Dante attributes to the divine emanation in Canto 18 of the *Paradiso* – which unfolds in the mathematical continuum and in the primordial generative impulse symbolized by the infinitesimal or differential – can be recognized in the sense of an infinite divine unfolding 'in time and action'. This theological perspective manages to perceive in every natural event a divine operation and

---

35. Gioberti, *Protologia*, p. 191.
36. Ibid., p. 159.

therefore an actual infinite. This same perspective helped to reverse the identification of 'cosmic evil' and 'infinite' in direct references to a positive infinity. If we look back to the Middle Ages, the first signs of this change are to be found in Duns Scotus' theses about Christ's infinitely expanded state of grace, in Gregory of Rimini's demonstrative arguments in favour of the categorematic infinite, and in Scholastic investigations into the 'breadth of forms' (*latitudo formarum*), or into the degrees of intensity, minimum and maximum, of qualities and forms.

But some of these ideas were already found among the Greeks. Gioberti refers to Melissus of Samos as the only one to intuit in *apeiron* a synonym of Being, and thus a positive unity, which at the same time he associated with the idea of the continuum. In a fragment which Aristotle refuted in his *Physics* (3.6.207a15), Melissus said of *apeiron*: 'It is therefore eternal and infinite and one and continuous (*homoion*).'[37]

---

37. See Renzo Vitali, *Melisso di Samo* (Urbino, 1973), p. 134.

4

The Infinite of St Thomas Aquinas

If we examine the definition of *set* that Georg Cantor proposed at the end of the nineteenth century, we cannot fail to be surprised by its evident resemblance to the definition which St Thomas Aquinas had formulated more than six centuries earlier.

For Cantor, a set is the gathering into a whole of objects determined and distinguished by our intuition or by our thought. For Aquinas, a set is likewise a collection of distinct units, that is: *'Multitudo non est aliud quam aggregatio unitatum.'*[1] And he further writes: *'Multitudo est id quod est ex unis quorum unum non est alterum'* (literally: 'A set is a composite of units, each one of which is distinct from the other').[2]

Needless to say, the apparent resemblance of these definitions in reality conceals completely different ideas and conceptions. Both definitions wish to capture the essence of what is supposed to be a set, but in Cantor's definition it is the urgency of a mathematical criterion that seems prior and predominant.

Cantor was motivated primarily by mathematical considerations in justifying the existence of sets that are actually infinite and can be associated with transfinite numbers. By contrast, for Aquinas every definition of set and number was simply a corollary of metaphysical rules, an arithmetical by-product of a vision of the world, more than a thousand years old, which lent Thomist theories no less verisimilitude

---

1. Cf. M. Périer, 'À propos du nombre infini', *Revue pratique d'apologétique*, 15 March 1919. [St Thomas literally writes that 'a multitude is merely an aggregate of units.' In the present chapter, I render Italian *insieme* by 'set', but 'aggregate' may be more appropriate to Thomist contexts. Translator's Note.]
2. Ibid.

than that seemingly possessed by Cantor's theories. Moreover, Cantor's notion of transfinite number was not born as a 'necessity', but rather as an 'opportunity', not as an obligatory deduction, but as the free creation and invention of this mathematical genius. As a matter of principle, one could question his introduction of set theory into mathematical language; and in fact some, like Kronecker, never tired of attacking Cantor with manifest hostility.

Cantor affirmed the existence of actually infinite sets and of infinite numbers. But six centuries earlier, Aquinas had authoritatively denied it. He argued that, if the elements of a supposedly infinite set were all simultaneously conceivable, so as to form an actually definite totality, they could be counted one by one. This would inevitably result in a finite number and therefore generate a contradiction.

But more than from their logical correctness, the force of these Thomist conclusions derived from a vision of the world which Cantor could but dimly perceive in his day. The incompatibility between the theses of Aquinas and Cantor is simply the incompatibility between the two different conceptions of the world that had generated them. Any possible identity between them makes very little difference, just as the formal identity of their definitions of set has no substantial importance.

It may be risky to dwell too long on the simplistic notion that Cantor was able to see what Aquinas had not grasped because the latter could have had no training in a branch of mathematics that did not yet exist. In this context, we may at least judge defensible the view expressed by Musil, who wrote that 'every advance is a gain in particular and a separation in general.'[3] Moreover, this opinion agrees with the more recent conclusions formulated by Thomas Kuhn in *The Structure of Scientific Revolutions*. The success of a new theory, or of a new scientific paradigm that bears the seeds of an authentic revolution, implies a total redefinition of all the corresponding sciences, and lends credence in the most general way to a new *Weltanschauung*. 'As the problems change, so, often, does the standard that distinguishes a real

---

3. Musil, *Man Without Qualities* II.40; Italian translation, pp. 146–7. [English version, trans. Wilkins, p. 163. Translator's Note.]

scientific solution from mere metaphysical speculation, word game, or mathematical play. The normal-scientific tradition that emerges from a scientific revolution is not only incompatible but often actually incommensurable with that which has gone before.'[4]

From this perspective, we may attribute to each thing a property that will one day place it in a great new correlation, without it necessarily constituting a conclusive truth. As Musil further writes, 'the mind or spirit is the great opportunist, impossible to pin down or take hold of, anywhere; one is tempted to believe that of all its influence nothing is left but decay.'[5] Not even the seeming incontrovertibility of the mathematical instruments used by Cantor could elude the equalizing justice of this rule. He simply knew how to find the just mathematical correlation into which the actual infinite could be incorporated as an idea that was both coherent and stimulating.

Still, it is worthwhile to talk about Aquinas, since even Cantor never shook off the feeling that contradicting Aristotelian and Thomist prohibitions exposed him to risks that were difficult to gauge. In his writings, he is constantly aware of the need to justify his conclusions in the light of the more traditional theories to which they ran counter.

In very clear and explicit terms, the *Summa theologica* alludes to the illogicality of God creating any object that is absolutely infinite (*infinitum simpliciter*). Aquinas writes: '*Contra rationem facti est quod sit simpliciter infinitum. Sicut ergo Deus licet habeat potentiam infinitam, non tamen potest facere aliquid non factum, hoc enim esset contradictoria esse simul, ita non potest facere aliquid infinitum simpliciter.*'[6]

The Thomist argument was thus simple and linear. God can make what He wants, and His making brings about the existence of what is

---

4. Thomas Kuhn, *The Structure of Scientific Revolutions* [2nd edn (Chicago, 1970), p. 103. Translator's Note]; Italian translation, *La struttura delle rivoluzioni scientifiche* (Turin, 1969), p. 132.

5. Musil, loc. cit.

6. Thomas Aquinas, *Summa theologica*, Part 1, Question 7, Article 2. ['It is against the nature of a made thing to be absolutely infinite. Therefore, as God, although He has infinite power, cannot make a thing to be not made (for this would imply that two contradictories are true at the same time), so likewise He cannot make anything to be absolutely infinite.' Translator's Note.]

made. Precisely because it is made, what is made cannot be completely without limits: '*contra rationem facti est quod sit simpliciter infinitum.*' Aquinas denied the existence of any absolute and actual infinite except for God. As proof of his thesis, he adduced an argument that in fact relied on the recognition of God's omnipotence. Hence, there is no contradiction between divine omnipotence and the impossibility of an absolutely (*simpliciter*) actual infinite.

The affirmation of this compatibility is important, because one of the first explicit ruptures with the Aristotelian–Thomist tradition sprang from precisely this point. In 1277, the bishop of Paris publicly condemned a considerable body of Averroist theses which implicitly struck at the Thomist thesis that God could not possibly create an infinite set. Later, some thinkers extended divine omnipotence to the creation of infinite quantities – a clear sign that *apeiron* was no longer considered synonymous with the negative component of the fundamental bipolarities of existence.

But we must proceed in orderly fashion if we are to understand clearly the sense of innovation that these later conceptions of the unlimited entailed. Above all, in what sense did Aquinas re-propose the Aristotelian notion of *apeiron*? Aristotle had excluded any possibility of confusing the false infinite of *apeiron* with infinite divine perfection, simply by denying the latter any infinite attributes. Divine perfection was designated by terms referring to its 'totality', its 'plenitude', and to its 'eternity', but not to its unlimitedness. The last term belonged exclusively to the realm of quantity, and was therefore completely extraneous to God.

Aquinas dares not follow Aristotle's formulation of the problem, and instead embraces the thesis that 'God is infinite and eternal and boundless'.[7] But he immediately adds that the infinite can have two opposite natures: one derived from the idea of form, and the other from the idea of matter.

The infinite 'on the part of matter' (*ex parte materiae*) thus had to correspond to an analogue of the false infinity of Aristotle's *apeiron*. By contrast, the infinite 'on the part of form' (*ex parte formae*), by

---

7. Ibid., Article 1.

referring to a sort of formal perfection, could indicate in what sense one could speak correctly of God's infinity.

This distinction between two infinities – which in essence resembled the distinction between potential and actual infinities – was thus a means of reconciling the Christian thesis of God's infinity, as proclaimed since the age of Tatian and Tertullian, with Aristotle's resolute assignment of the infinite to the material principle of existence. The one mirrored the imperfection of *apeiron*; the other, God's perfection. Aquinas writes: '*Infinitum secundum quod attribuitur materiae, habet rationem imperfecti, est enim quasi materia non habens formam,*' but '*infinitum secundum quod se tenet ex parte formae non determinatae per materiam, habet rationem perfecti.*'[8]

Yet in the concrete world of becoming, neither the *infinitum ex parte materiae* nor the *infinitum ex parte formae* can be found in its bare essence. The former cannot exist simply because every unlimited potentiality is modified by a boundary imposed on it by the formal principle '*materia autem perficitur per formam per quam finitur.*' And it is difficult to rank the latter among created forms, since form is in a certain sense 'contracted' by the material that causes it to exist as the essence of a particular individuality. Thus, if form is granted unlimited expansion in conformity with its degree of universality, it ceases to constitute the essential attribute of one object or another, and so becomes identified with what has no need of external receptiveness to exist, because it finds its own substantiality within itself. The infinite *ex parte formae* (on the part of form) and therefore, in sum, true actual infinity are typically identified with the divine essence: '*cum igitur esse divinum non sit esse receptum in aliquo, sed ipses it suum esse subsistens . . . manifestum est quod ipse Deus est infinitus et perfectus.*'[9]

So can any infinity exist, except for God? To this crucial question, Aquinas responds by leaving open a faint possibility of existence. He

---

8. Ibid. ['The infinite as attributed to matter has the nature of something imperfect; for it is, as it were, formless matter . . . the infinite regarded on the part of form not determined by matter has the nature of something perfect.' Translator's Note.]

9. Ibid. ['Since therefore the divine being is not a being received in anything, but He is His own subsistent being . . . it is clear that God Himself is infinite and perfect.' Translator's Note.]

concludes that absolute infinity *per essentiam* or *simpliciter* pertains only to God, while what is different from Him can only be finite (*finitus simpliciter*) or infinite in a relative way, that is, corresponding to its specific nature (*infinitus secundum quid*). Therefore, if the existence of the absolutely (*simpliciter*) infinite is completely precluded in the world of forms, the *infinitum secundum quid* is still possible – which leaves room for a reasonable doubt that there might possibly be an exception to Aristotle's rigorous rejection of any kind of infinity. But Aquinas seems to imply that the *infinitum secundum quid*, as considered in the realm of human comprehension, can be essentially identified with the *infinitum ex parte materiae*, and is therefore a potential (i.e., non-existent) infinite.

It is easy to cite a familiar instance of the *infinitum secundum quid* that partakes of the material principle. Aquinas adduces the example of wood which, while finite in its essence, is infinite if viewed from the point of its unlimited potentiality for being formed into infinite figures. In turn, the *infinitum secundum quid* that partakes of the formal principle can no longer be easily associated with an object of ordinary experience. This would mean displaying universal forms not 'contracted' by matter, as we might imagine only angelic essences could exist. As Dante would later write, in the Angels 'pure act was produced'.[10]

If at this point we wish to grasp the nature of the quantitative infinite, which for Aquinas is also the infinite studied by the mathematicians, we need only add a simple corollary to our previous account. First of all, the infinite *secundum quantitatem* is clearly associated with the potential infinite, and with the infinite *ex parte materiae*, and thus is not attributable to God: '. . . *infinitum quod competit quantitati, est infinitum quod se tenet ex parte materiae. Et tale infinitum non attribuitur Deo.*'[11] The infinite quantity (*quod competit quantitati*) thus transfers the negative character of its own potentiality to the mathematical infinite, which was intimately linked to the notion of quantity. So when Aquinas refers to the mathematicians' use of the infinite, he repeats

---

10. Dante, *Paradiso* 29.33.
11. Thomas Aquinas, *Summa theologica*, Part 1, Question 7, Article 1. ['The infinite of quantity is the infinite of matter; such a kind of infinite cannot be attributed to God.' Translator's Note.]

what Aristotle had already asserted. In order to prove his own theorems, the geometrician has no need of the actual infinite. He has only to verify that it is possible, in certain cases, to proceed beyond any given magnitude or finite quantity, so that he may safely use the infinite in its limited and only permissible meaning, as potential: '*geometer non indiget sumere aliquam lineam esse infinitam actu; sed indiget accipere aliquam lineam esse finitam actu, a qua possit subtrahi quantum necesse est.*'[12]

Even while existing in perfect forms, the infinite *secundum quid* and the infinite *ex parte formae* thus find no place in the world of mathematical forms, where the ancient distinction between limiting form and unlimited matter remains in effect. The formal character of mathematics relies on the limits of geometric figures and on the ordering perfection of finite numbers.

Just as for the Greeks, for Aquinas too the surface outlines of objects and their geometric figure were the terminus and boundary *par excellence* of the unlimited potentiality of every material substratum, as well as the miraculous point of encounter and equilibrium of this potentiality with the 'contraction', effected by matter, of God's perfect form in the limited forms of existence.

This truth had been described by Euclidean and Pythagorean mathematics, in which the colour of an object was added to its surface outline to confirm the limit of its existence. The Pythagoreans called the surface of bodies *chroia*, or colour.[13] Much later, this correspondence was also to be emphasized by Goethe: 'The eye sees no form, since only light, dark, and colour *together* make up what for the eye distinguishes one object from another, or one part of an object from another.'[14]

In fact, even if we grant the existence of 'scotoptic vision' (the visual perception of objects in the dark), which seems to contradict Goethe's thesis, it would still be difficult to imagine that form, and hence limit,

---

12. Ibid., Article 3. ['A geometrician does not need to assume a line actually infinite, but takes some actually finite line, from which he subtracts whatever he finds necessary.' Translator's Note.]

13. T. L. Heath, *A History of Greek Mathematics* (Oxford, 1921), vol. 1, pp. 292–3.

14. Goethe, *Theory of Colours*; Italian translation in *Opere*, vol. 5 (Florence, 1961), p. 298.

would retain *all* their semantic fullness and universality without colour. If we turn back to antiquity (and thus to Goethe as well), it is impossible to discuss 'form' without invoking its symbolic function as a link between different planes of existence, or as a class of actual equivalences that correspond mutually and can be assimilated to a single sign. If we left out colours and their function as outline and surface, we would lose an entire tradition of correspondences; and it would be difficult to rethink the set of attributes traditionally associated with form in this unusual and spectral guise. Merely reading the classic essay of Frédéric Portal, *Des Couleurs symboliques* (*On Symbolic Colours*) (Paris, 1837), will convince us that we could not reject the Greek identification of *peras* (limit) and *chroas* (colour) without sacrificing the larger significance of the cultural tradition that presumably inspired it. Besides, as Goethe wrote, 'The metaphysics of the theory of nature should remain the field of the philosopher, who should be allowed to decide at what height or depth it will begin and how far it will ascend or descend.'[15] This statement leads us directly back to Kuhn (and, more obviously, to Feyerabend) and to the reciprocal incommensurability of paradigms: Goethe's theories clash on the factual level with observed data, but fit into the inner compatibility of a schema that could not have reconciled *eidos* (form) with darkness without contradicting itself.

Aristotle too explained that the chromatic quality of bodies was inseparable from their limit (*peras*), and hence from their geometrical outline. And the identity of the limit and the surface of a solid was synthetically described by the second definition in Book 11 of Euclid's *Elements*: 'The limit of a solid is its surface.'

Even the Aristotelian definition of humidity was based on identical principles. What is humid shares in the unlimited because it lacks its own autonomous surface outline: this is why water assumes the form of the receptacle that contains it.[16]

Analogous concepts are evident in the first sentence that Aquinas

---

15. Goethe, *Teoria della natura*, ed. M. Montinari (Turin, 1968), p. 298.
16. Aristotle, *On Generation and Corruption* 2.2.329b30. See also E. Zolla, *Le meraviglie della natura. Introduzione all'alchimia* (Milan, 1975).

writes in countering the use of the infinite in mathematical procedures: '*Omne corpus superficiem habet. Sed omne corpus superficiem habens est finitum, qui superficies est terminus corporis finiti. Ergo omne corpus est finitum. Et similiter potest dici de superficie et linea. Nihil est ergo infinitum secundum magnitudinem.*'[17]

Even when he maintains that the infinite is unknown, no matter what its species, Aquinas reaffirms a Pythagorean truth.[18] In placing the infinite on a par with the material principle, the Pythagoreans had declared it unknowable. To the unknowableness of the infinite in magnitude and multitude (*secundum magnitudinem et multitudinem*, or the mathematical infinite *ex parte materiae*), Aquinas adds the inscrutability of the infinite in species or in quality (*secundum speciem* or *secundum qualitatem*), and finally the inscrutability of the divine being. Dante was to imitate him in his *Convivio* when he wrote that a property of the Sun is reflected by the reference in arithmetical laws to a numerical infinite: 'The second property of the Sun we also see in number, of which Arithmetic is the science: for the eye of the intellect cannot see it, since number, considered in itself, is infinite; and we cannot understand this.'[19]

From the impossibility of an actually infinite set, Thomas derived the finiteness of the unit by which the set was measured, namely, of number.[20] Once a set had been defined as a collection of distinct units, number corresponded to a further intellectual operation that endowed it with measure, and thus with order, by further specifying its formal

17. Thomas Aquinas, *Summa theologica*, Part 1, Question 7, Article 3. ['Every body has a surface. But every body which has a surface is finite; because surface is the term of a finite body. Therefore all bodies are finite. The same applies both to surface and to a line. Therefore nothing is infinite in magnitude.' Translator's Note.]

18. Thomas Aquinas, *Commentary on Aristotle's Physics* 1.1, lect. 9.

19. Dante, *Convivio* 2.13.19, in *Opere* (Florence, 1968), part 1, p. 202 ['L'altra proprietade del Sole ancor si vede nel numero, del quale è l'Arismetrica: che l'occhio de lo 'ntelletto no può mirare; però che 'l numero, quant'è in sé considerato, è infinito, e questo nol potemo noi intendere.' Translator's Note.]

20. St Thomas Aquinas distinguished between the infinite in multitude (*secundum multitudinem*) and the infinite in magnitude (*secundum magnitudinem*). The actual non-existence of the former is discussed in *Summa theologica*, Part 1, Question 7, Article 4.

character of being knowable as a limited whole. The set is counted *per unum*, i.e., by numbering one by one each of the indivisible units of which it is composed. The last unit counted determines the final result, which is constituted by a numbered amount that is necessarily finite. Thus, the final number in a way seals the existence of the set as an actual totality, assigning it a concrete existence on the basis of the effective and simultaneous existence of all the numbered elements.

Here are the Thomist definitions: 'Number is a plurality measured by one . . . Any number is a multitude measured by one . . . Number adds to a multitude the reason of measurement.'[21]

In support of his argument, St Thomas's *Summa theologica* cites the Book of Wisdom 11.20, which ascribes to God's creation a geometrical division of the cosmos: 'You have ordered all things by measure and number and weight.' And he adds that he thinks that 'everything created must be necessarily comprehended in a certain number': *'Necesse est quod sub certo numero omnia creata comprehendantur.'*[22] We could also add, without misinterpreting his meaning, that the infinite coincides in a Pythagorean manner with the non-measured, non-created part of the cosmos, and is therefore even less compatible with number. Thus, by associating number with the concept of measure (conceived almost Platonically as the criterion of the demiurgical creation of the world), Aquinas was led to conceive of infinite number as a contradictory notion – senseless, useless and misleading for anyone who sought to discern God's unfathomable perfection in the perfection of limited forms.

21. ['*Numerus est pluralitas mensurata uno . . . Quilibet numerus est multitudo mensurata per unum . . . Addit enim numerus super multitudinem rationem mensurationis.*' Translator's Note.] See M. Périer, 'À propos du nombre infini.'
22. Thomas Aquinas, *Summa theologica*, Part 1, Question 7, Article 4.

5

Categorematic Infinite and
Syncategorematic Infinite

Medieval thinkers invented and debated an original formula that distinguished between the concepts implicit in the actual and potential infinite. In the seventh treatise of his *Summulae logicales*, Peter of Spain – who is generally identified with Pedro Julião (1226–77), later Pope John XXI – teaches that the infinite can be understood in two distinct ways. The first, denoted by the term 'categorematic', sought to express something that resembled the actual infinite without being identified with it. The second, denoted by the term 'syncategorematic', sought to express the idea contained in the potential infinite while depriving it of the principal ambiguity implied by the use of the term 'potentiality'.

The difficulty inherent in the Aristotelian association of *apeiron* and *dynamis* was the following: potentiality (*dynamis*) always presupposes an end towards which it is directed, and therefore cannot exist without an act. (And a potentiality, Averroes commented, is defined by the very term for its action, and not by what precedes this term.)[1] By contrast, *apeiron* indicates a condition contrary to actuality; and Aristotle specified that the unlimited, understood as something larger than any finite magnitude, could not exist even potentially.[2] Thus, to speak of a 'potential infinite' might generate confusion: the risk was that the term 'infinity' anticipated a presumable object towards which its potentiality was oriented. Such an operation was not always legitimate, since the indefinite character of the *apeiron* is typically manifested by its

---

1. On Averroes' comment, see P. Duhem, *Le système du monde*, vol. 7, p. 72.
2. Aristotle, *Physics* 3.6.206b20.

'unconditioned' opening to every form of possible event, and hence to absolute disorder, to pure chance.

In the syncategorematic infinite and in the formulas devised to denote it, every reference to potentiality was eliminated, and hence every reference to the possibility of actual realization that it implies. As the concept was later defined by Gregory of Rimini and Buridan, the 'open' nature of the syncategorematic infinite, as well as the repeatability of the pure finite in which it consists, could be simply characterized in the following case: given a finite quantity, however great, there exists an even larger quantity. One may also affirm that even the finite body is infinite in the syncategorematic sense, since to every finite body there corresponds a larger body: *'infinitum est corpus finitum, quia omni corpore esset maius corpus finitum.'*

By contrast, the categorematic infinite was the attribute of a subject posited as 'something' larger than any existing finite magnitude. Gregory of Rimini taught that one could define the categorematic infinity by merely transposing the words that denote the syncategorematic infinite. From *quantocunque finito maius*, a phrase that expresses the syncategorematic by emphasizing an increasable finite, one derives *maius quantocunque finito*, a phrase that implies a determinate subject preceding its own attribute of infinity.[3]

One of the most debated examples was that of the *linea gyrativa* (spiralling line), which Buridan characterized as an extremely difficult problem. Let us suppose that there is a cylinder with a unitary height h, and let us divide it into 'proportional' parts, that is, repeatedly subdivide its height by half ($\frac{1}{2}$, $\frac{1}{4}$, $\frac{1}{8}$, . . .) to create an infinite series of cylinders. (This is the same way in which Zeno divided a unitary distance in his first argument against motion.) On the surface of the first partial cylinder, we draw a spiral with a 'pitch' equal to its height. Then we extend it on to the second partial cylinder, which has a height of $\frac{1}{4} = \frac{1}{2}^2$, and add another stretch of the spiral with a pitch of $\frac{1}{4}$. On the nth partial cylinder ($n = 3, 4, 5, . . .$), we extend the *linea gyrativa*, which has reached the ($n - 1$)th cylinder

---

3. Duhem, *Le système du monde*, vol. 7, p. 63. [Duhem notes that this is untranslatable. Translator's Note.]

through decreasing pitches, and add the nth stretch of the spiral, with a pitch of $\frac{1}{2}^n$.

In the end, we may ask ourselves what kind of infinite has been realized by the various pitches of spiral that traverse the surface of the hypothetical cylinder. If we believe that the entire cylinder, conceived as an actually given whole, determines the existence of the *linea gyrativa* as a categorematic infinite, we shall presumably be authorized to affirm that 'the *linea gyrativa* is drawn across all the parts of the cylinder.' But if we attribute to the line a merely syncategorematic infinity, we may rather say that 'across every part of the cylinder is drawn a *linea gyrativa*.' Thus, a simple verbal permutation serves to shift between two opposing viewpoints in visualizing this elusive unlimited quantity. The first posits an actual limit from which we may contemplate the indefinite potentiality of the series that tends towards it. The second focuses on the incessant nature of the additive process, rather than on an ideal point of arrival.

It was observed that, whenever the syncategorematic infinite became the attribute of a subject, it deprived the subject of any sign of actuality and integrity. Not that the syncategorematic infinite could not in fact refer to a subject, Peter of Spain pointed out, but by such an attribution the subject's individuality was automatically destroyed, and its precise contours lost. For example, in the sentence 'An infinity of men are running,' the word 'men' merely denotes a confused multitude, rather than a set of determinate individuals; and even the addition of two or three individuals cannot change the meaning of the sentence.

The difficulty that Buridan noted in the case of the *linea gyrativa* probably lay in its cryptic superimposing of the two opposite natures of the unlimited: (1) the indefinite-by-addition of the various finite pitches of the spiral, which is typically syncategorematic and lacks any intrinsic solution; and (2) the infinite-by-division of the proportional parts of the entire cylinder which – being posited as a whole *a priori*, i.e., before the division – could confer on the aggregate of its parts the attribute of categorematic infinity.

We are not interested here in exploring the more arcane reasons behind such difficult problems, nor in expounding the subtleties of the logical arguments that were devised to resolve them. Pierre Duhem has described them exhaustively in every detail, offering an ample survey of the cases involving the problems and examples which, like the *linea gyrativa*, exercised the logicians of Oxford and Paris.

Yet there is an important aspect of those discussions which deserves to be mentioned. It lies in a sort of correspondence and contamination, often dimly adumbrated, between the infinite power of God and the indefiniteness of *apeiron*. Logical arguments often called upon divine intervention in the spectacle of the eternal absurdities which the destructive power of the false infinite caused in the reality of forms. And this divine role accompanied the incessant stages of the false infinite, which according to the logic of becoming sometimes tended towards the imaginary resolution of the form *par excellence*, the actual infinite. In this way, the unlimited nature of *apeiron* began to appear as a plausible sign of the immanent operation of God's unlimited nature in the world. This confusion was dangerous, because God's infinite and the infinite of *apeiron* had to retain their opposite natures, when considered from the (illusory) perspective of the fundamental dualism between good and evil, between limit and the unlimited, and between positive and negative.

One might object that God's power is divine for precisely this reason: whenever it encounters something unlimited, it resolves it into formal unity, and causes it to exist as an actual infinite. This makes its operation legitimate rather than contradictory. For by acting in this way, the divine does not mingle with the perplexities of the false infinite, but instead eliminates them by its resolving action. Yet in reality it did not always happen that formal unity was clearly established when the *deus ex machina* of God's omnipotence intervened. Sometimes the purely unlimited extension of magnitude was emphasized; whereas scant attention was paid to the miracle of form, which envelops every existing thing, regardless of its magnitude, and even in spite of it.

For example, Gregory of Rimini divided a unit of time, an hour, into proportional parts (*partes proportionales*): $\frac{1}{2}, \frac{1}{4}, \frac{1}{8}, \frac{1}{16} \ldots \frac{1}{2}^n \ldots$ exactly as Zeno had repeatedly halved a unitary segment. Next, Gregory imagined that God could create a stone during each portion of time, and add it to the previous ones. It is obvious, he concluded, that at the end of an hour God will have created an infinitely large stone.

The example is basically similar to that of the *linea gyrativa*: for both assign to each 'proportional' part an Archimedean quantity which, being added to itself an infinite number of times, generates a quantity larger than any given finite quantity. But the two examples are also dissimilar. In the case of the *linea gyrativa*, the description is purely geometrical. By contrast, the creation of a stone during each proportional part of an hour requires God's explicit and direct intervention. And this at least temporarily violates the laws of creation, which require the contraction of God's infinity within limits of form, or a correction, moment by moment, of the indefiniteness of *apeiron* through the precise outlines of a given figure.

Gregory of Rimini's 'proof' obviously claimed that God's power could reveal itself in the absurdity of infinitely expanded matter, and that the actual infinite would be realized in the tangibility of such a miracle. Although Zeno had discussed a similar example, his distance from Gregory of Rimini is enormous. Zeno had allowed the disintegrating power of the unlimited to destroy the limits revealed by sensory experience and by rational corollaries: to this end, the negativity of *apeiron* was indispensable. By so doing, he had revealed the paralysis that lay behind an illusory reality governed by the reciprocal contrast between limit and the unlimited. For Gregory of Rimini, the negativity of *apeiron* no longer serves a purpose. Instead, this negativity is changed into its opposite: to wit, into the positive affirmation of unlimited growth, which itself is capable of achieving perfection in an immanent form. The negative theology of Zeno's *apeiron* thus crosses over into sensory evidence. What Gregory of Rimini obtains is an infinitely *large* stone, and this connotation of magnitude is the emblematic sign that the mystery of the infinite has been rendered external.

What, then, is wrong with something being or becoming *magnitude*? In his *Enneads*, Plotinus answered this question by observing that 'a thing exists more, not when it comes to be many or large, but when it belongs to itself,' that is, 'in tending to itself', into its own 'interior self'. 'Through magnitude and as far as it depends on magnitude, it loses itself,' he wrote; 'but as far as it possesses unity, it possesses itself.'[4] For

---

4. Plotinus, *Enneads* 6.6.1; Italian translation (Bari, 1973), pp. 286–7.

Plotinus, the Universe was beautiful *despite* its magnitude; it was its constant embrace with the One that kept its flight towards the unlimited from degenerating into mere dissipation.

This rupture of limit could be viewed in two intimately connected senses. The most obvious sense – the inordinate growth of a magnitude (Gregory of Rimini's stone) – presupposed God's ability to overcome the minimal threshold of time required by every finite interval of becoming. (It was presumed that in the creative act God was not obliged to respect the Archimedean minimum as necessary for the existence of any form, or for any finite action.) Obviously, once exceeding the minimal threshold was declared permissible, this entailed the possibility of the infinitely large, which became a symbol of divine omnipotence. And it is tempting to suppose that after this point limit began to represent the negative counterpart of such omnipotence, as an obstacle to its unfolding: the most obvious of arguments insists that the negative opposite of God's infinity is finitude or limit.

We could cite other examples. Nicolas Bonet, who stoutly defended the possibility of the actual infinite, maintained that 'the productive force of the Prime Mover extends to the production of numerical infinites' (*secundum multitudinem*).[5] Nor was God's power limited to the creation of a relative actual infinite (*infinitum secundum quid*); it could doubtless extend to absolute actual infinites (*infinitum simpliciter*).

Many of the arguments current in the schools of Oxford and Paris, both for and against the actual infinite, were later collected by Jean Mair in his *Propositum de infinito*, written in the early 1500s. In it, he particularly stresses the fact that God can produce a body of indefinitely increasing magnitude. For otherwise, there must absurdly exist a finite body which God could not even extend by a foot: '*quocunque corpore finito creato Deus potest creare maius, alioquin oportet dare unum corpus finitum cui Deus non potest addere unum corpus pedale.*'[6]

The assertion is rather obvious, since it is fairly natural not to

---

5. Duhem, *Le système du monde*, vol. 7, pp. 129–30.
6. Jean Mair, *Le traité 'De l'infini'*, ed. H. Élie (Paris, 1938), p. 92. ['God can create something larger than any finite created body; otherwise we must grant that there is a finite body to which God cannot add even a foot-long body.' Translator's Note.]

deny the existence of a syncategorematically infinite magnitude. Less obvious and more significant is the insistence on God's absolute and perfect omnipotence in order to justify the validity of the thesis. And even less obvious is the invoking of God's omnipotence to defend the credibility of a categorematically infinite magnitude, or of an aggregate, infinite in the categorematic sense, comprising objects that are separate and lacking in continuity.

Jean Mair followed the established practice of deriving the syncategorematic infinite from the categorematic infinite, and vice versa, by means of a simple exchange of words, when he wrote: *'Prima veritas: infinite magnum corpus Deus potest producere. Secunda veritas: Deus potest producere corpus infinite magnum. Tertia: Deus potest producere infinitum multitudine rerum separatarum nec continuo inherentium.'*[7]

To bolster this lapidary confutation of Aristotelian and Thomist theses, Jean Mair used arguments not unlike those already employed by Gregory of Rimini and by others who had defended the existence of the actual infinite, including the possible creation of a stone in any 'proportional part' of an hour.

Still, we should note that, if an actual infinite was compatible with the created world, this did not automatically contradict the underlying reasons that suggested the opposite thesis in Greco-Thomist thought: namely, that infinity could exist as pure potentiality.

Like Gregory of Rimini, Jean Mair also tried to describe the perfection of form in a way not very different from those who conceived of it as the limit of an infinity. When he maintains that every body possessing an infinite set of parts (i.e., proportional parts) is an actual infinite (*'omne corpus habens infinitas partes est infinitum actu'*), he is merely referring – albeit with dubious logic – to a truth that is by no means improbable or inconsistent with traditional views.[8] The Pythagoreans as well, who affirmed that everything contains both limit and the unlimited, could have admitted that every body can be

---

7. Ibid., p. 90. ['First truth: an infinitely large body God can produce. Second truth: God can produce a body infinitely large. Third: God can produce an infinite in multitude of things separate and not bound in a continuum.' Translator's Note.]
8. Ibid., p. 8.

represented both as an indefinite set of proportional parts and as the actual totality that contains these parts. As for Aristotle, while he rejected the actual infinite, he too alluded to the possible limitation of processes carried to infinity.[9] And by the same token, Plotinus had implied that a limit can only be the limit of an infinity.[10]

In the angel's greeting *Ave gratia plena* ('Hail, full of grace'), Jean Mair saw a direct allusion to the perfect form as an actual infinity, a point of reference and comparison for man's partial, imperfect and necessarily finite participation in God's supreme grace.[11]

Duns Scotus had anticipated Jean Mair by an analogous argument even if he did not refer explicitly to the logical consistency of an actual infinite.[12] Arguing about grace and whether it can be increased indefinitely, he arrived at the conclusion that there exists a perfect grace which is not susceptible of further increments. And the same is true of every form to which one can attribute an increasing or decreasing intensity, such as colours. Perfect white is, in a certain sense, of infinite intensity, for it appears as a point at infinity in an unlimited series of approximate gradations that tend to achieve the definitive form. The reference to Aristotle is explicit: 'What can exist in act is the measure of what can exist in potentiality.' Hence, for every form one may inevitably hypothesize an insuperable limit that simultaneously subsists for all intermediate degrees, such that it also constitutes the principal criterion of their hierarchy. The reference to the Aristotelian priority of act is here employed to undermine the Aristotelian thesis that the infinite does not exist. Act and infinite are nearly in harmony.

Yet the defence of the actual infinite was not the only reason for breaking with tradition. In reality, even to those who followed Aristotle or St Thomas in defending the sole existence of the syncategorematic infinite, the boundary by which the world had been previously delimited was now clearly and obviously transcended. Indeed, the sense of the innovation was even more transparent when one stopped viewing the 'open' nature of the syncategorematic infinite as the negative

---

9. Aristotle, *Metaphysics* 2.2.994b9.
10. Plotinus, *Enneads* 6.6.3; Italian translation (Bari, 1973), p. 288.
11. Jean Mair, *Le traité*, p. 102.
12. Duns Scotus, *Sentences*, Book 3, dist. 13 (ed. Durand, Lyons, 1639), vol. VII.1.

counterpart of the perfectly 'limited' world created by God, and used it as a positive receptacle of the progressive unfolding of divine power. In the traditional view, the world could only be contemplated *sub specie aeternitatis* (in its eternal aspect) to the extent that it could be intuited as a limited totality. (Wittgenstein reiterated this in his *Tractatus*.) But this very idea was now reversed in the image of a Universe in which the sign of God was marked by its unlimited expansion.

A thirteenth-century Franciscan, Richard of Middleton, was one of the first to admit that the Universe can expand beyond any given limit, without it necessarily implying the notion of an actually infinite world. Consequently, he also asserted the possibility of a plurality of worlds and the possibility that God could impart a motion of translation to the heaven of fixed stars, which had traditionally been regarded as immobile.[13]

Richard of Middleton writes: 'God can produce a magnitude or a dimension that increases without end, provided that at each moment the magnitude actually realized is finite. In the same way, God can indefinitely divide a continuum in parts whose magnitude falls below any limit, provided that there never actually exists an infinite number of really divided parts.'[14]

Gilson calls this thesis 'the unexpected result' of the condemnation of Arabic Aristotelianism, which Étienne Tempier, bishop of Paris, issued in 1277. Pierre Duhem does not hesitate to classify this event as the birth certificate of modern science. And Gilson, while more cautious, still regards it as the first indication that it was now possible, in a Christian context, to conceive the world in keeping with modern cosmological theories.

The Parisian condemnation of certain propositions also clearly determined the use that future mathematicians would make of the infinite. The birth of modern cosmology coincided with the first infractions of Aristotle's prohibition against using infinites and infinitesimals in arithmetic and geometric proofs.

---

13. See Étienne Gilson, *La philosophie au Moyen-Âge* (Paris, 1952); Italian translation, *La filosofia nel medioevo* (Florence, 1973), pp. 555–6.
14. Ibid., p. 555.

Article 29 of the condemned propositions has a particular importance, since it implies that God can create infinities and bring them into existence in the world of becoming and change. Literally, the Averroist thesis says that God's infinite can only be compatible with its total separation from the becoming of the Universe, and that every immanent manifestation of such infinity can only be found in the perfection of form: '*Quod Deus est infinitae virtutis in duratione, non in actione, quia talis infinitas non est nisi in corpore finito, si esset.*'[15]

At any rate, the thesis was associated with a typical aspect of Averroist doctrine, which placed the existence of intermediate causes between the absolute immobility of God and the mutability of created things. Indeed, another of the condemned propositions declared that 'the First Principle can be the cause on earth of different effects only by means of another cause, since nothing that changes things can effect changes of different kinds without being itself changed'.[16]

What did the condemnation of these propositions imply? First of all, it officially declared that God's 'direct' intervention in the world was compatible with the world's finite and limited nature. Secondly, certain incautious affirmations about the possible unfolding of divine infinity in time and act (*in tempore et actione*) could well have encouraged rash and unpredictable innovations, which risked compromising the finite and Archimedean nature of forms. This made possible the emergence of theses and proofs about the presence of both the actual and potential infinite in the world, now regarded as the positive unfolding of the transcendent infinite, so that *apeiron* was no longer synonymous with metaphysical evil, with absurdity, with non-being and (owing to its antithetical character) with a secret allusion to the hidden God (*Deus absconditus*). As Erwin Panofsky suggested, with reference to the stylistic changes in the representation of space dating back to Romanesque and Gothic art, even the emergence of Scholastic theses on God's infinity was in some way the prelude to an authentic 'actual

---

15. See Bruno Nardi, *Saggi sull'aristotelismo padovano dal secolo XIV al XVI* (Florence, 1958), p. 185. ['God is of infinite power in duration, not in action, since such infinity only exists in a finite body, if it indeed exists.' Translator's Note.]
16. See Gilson, *La filosofia nel medioevo*, p. 671.

infinite' (*energeia apeiron*) transferred from the supernatural sphere to the natural one.[17]

And if we wished to discover the source of our irresistible urge to investigate the infinite, we might well turn our attention to what Arthur O. Lovejoy called the 'principle of plenitude'. In *The Great Chain of Being*, Lovejoy says that this principle secretly governs much of the history of Western thought, and explains that it refers to the universe as a plenum of forms (*plenum formarum*), in which all the possible variants of *genera* of living things are exhaustively exemplified. This principle is also derived 'from the assumption that no genuine potentiality of being can remain unfulfilled, that the extent and abundance of the creation must be as great as the possibility of existence and commensurate with the productive capacity of a "perfect" and inexhaustible Source, and that the world is the better, the more things it contains.'[18] Lovejoy identifies traces of this urgent need – which seeks to fill in every empty interstice of reality and every glimmer of possibility – as far back as Plato's dialogues and Aristotle's treatises. And when we read many later affirmations that the actual infinite exists, might we not describe their content as in fact an *extrapolation* to infinity of this very principle?

---

17. See Erwin Panofsky, *Die Perspektive als symbolische Form*; Italian translation, *La prospettiva come forma simbolica* (Milan, 1976), p. 55. [English version, *Perspective as Symbolic Form*, trans. Christopher S. Wood (New York, 1991), p. 44. Translator's Note.]

18. Arthur O. Lovejoy, *The Great Chain of Being* (Cambridge, Mass., 1936), p. 52; Italian translation, *La grande catena dell'Essere* (Milan, 1966), p. 57.

6

Giordano Bruno, Nicholas of Cusa, Raymond Lull

We would be correct if we said that the Renaissance reintroduced infinitesimals and non-Archimedean quantities into mathematical applications, and that this represented a gain in terms of innovation but a loss in the formal correctness of proofs. But instead of grasping the larger meaning of this truth, we would only extract a fragmentary portion of it, the one that seems best suited to fit the history of a mathematics continually inspired by the urgent need for a rigorous linguistic description of its own concepts and its own free constructions.

In reality, it may happen that by its very nature the logical untenability of a mathematical discourse contains the seed of a truth, or that such an illogical quality can even become a hallmark of the only visible indication of inaccessible ideas or realities.

Simone Weil was not far from the truth when she wrote that 'mathematical invention is transcendent' and that 'it proceeds from absolutely non-representable analogies, and all we can do is to mark their consequences.'[1] It is quite remarkable how this viewpoint, which is so clearly alien to the canonical rules of modern mathematical practice, accords with the critique that Brouwer directed against the logicist orientation of mathematics. The first act of mathematical intuitionism is to free mathematics entirely from the language that expresses it, and to make it an autonomous mental activity. 'In the construction of mathematical thought . . . language only plays the role of a technique,

---

1. Simone Weil, *Cahiers*, vol. 2 (Paris, 1972), p. 156. [*Notebooks*, trans. Wills, p. 260. Translator's Note.]

efficient but never infallible or exact, for memorizing mathematical constructions and for suggesting them to others.'[2] In other words, in a sense that might be defined as 'absolute', the more recondite mechanisms of mathematical invention cannot be represented by language.

Thus, the existence of a discursively elusive something (*quid*) can naturally be offset by resorting to a descriptive science based on paradox, on apparent incongruence, and on an enigma mathematically expressible by configurations that defy any rational rigour. This is the art of 'learned ignorance' espoused by Nicholas of Cusa, as well as in Giordano Bruno's ultimate justification of the theory of geometric minima.

To represent the infinite mathematically, these men preferred a language that clearly displayed its peculiarities, even though this involved formal inaccuracies incompatible with pure rationality. Let us cite only a few examples. Raymond Lull suggested that mathematics should also embrace the task of describing concepts belonging to the sphere of intelligibility, and not only to the sphere of rationality.[3] Averroes declared that mathematics should be regarded as a discipline aimed at the study and comprehension of formal causes, and hence of archetypes.[4] Giordano Bruno profoundly respected Plato's intention of making the study of numbers and geometrical forms the link between the sensible world and the intelligible world of ideas. Bruno situated mathematics between physics and metaphysics.[5]

The geometric constructions of Raymond Lull, and especially those of Nicholas of Cusa, give rise to antinomies that form a continual reference to the ineffable absolute. The illogical and the irrational – whatever cannot be concretely drawn and seen on a piece of paper – are intentionally cited as entities whose nature and essence can only be grasped by the intellect, or '*mens tuens*': the intellect 'sees' what the eye cannot distinguish and the hand cannot draw.

2. L. E. J. Brouwer, 'Historical Background, Principles, and Methods of Intuitionism', *South African Journal of Science* 49 (1952): 139–43; Italian translation in C. Cellucci, *La filosofia della matematica* (Bari, 1967), pp. 223–31.

3. See Giordano Bruno, *Praelectiones geometricae et Ars deformationum*, ed. G. Aquilecchia (Rome, 1964), p. xx.

4. Ibid., p. xix.

5. Ibid., p. xxi.

Giordano Bruno expounded the same thing. He understood that, even if the antinomy is a revelation of the true actual infinite, the latter is necessarily incompatible with the world of forms. If we can speak of the infinite in this world, we inevitably deal with the unlimited, with the potential infinite. 'Of the divine substance,' he wrote, 'because it is both infinite and extremely remote from those effects which constitute the outer limit of the path of our discursive faculty, we can know nothing – except by means of vestiges, as the Platonists say; or of remote effects, as the Peripatetics say; or by means of garments, as the Cabalists say; or of dorsal and back parts, as the Talmudists say; or of a mirror, shadow, and enigma as the Apocalyptics claim.'[6]

The invisible point around which our discursivity strives to repeat and reformulate its partial verities, that is, the unattainable actual infinite, forces every other infinite that is accessible to the senses or to pure rationality into a fatal participation with potentiality and matter. Thus, inevitably 'our intellective potency is unable to comprehend the infinite, except in speech or in a certain manner of speaking, or, in other words, by a certain potential reason and natural disposition, and he of whom we speak does not differ from one who would aspire towards the immeasurable as an end where in fact there is no end.'[7]

Some scholars have written that Bruno confused true infinity with the indeterminacy of the false infinite. In fact, the contrary is true, as Bertrando Spaventa showed in an essay that also contains a lucid account of the meanings of the infinite.[8]

For Bruno, degrees of knowledge are arranged in a succession of increasing perfectibility which begins with the senses, and is progressively refined in our reason, in the intellect and finally in the mind. Rational discursivity merely adds the finite to the finite, struggling

6. Giordano Bruno, *De la causa, principio et uno*, ed. G. Aquilecchia (Turin, 1973), p. 63. [English version, *Cause, Principle and Unity*, trans. Robert De Lucca (Cambridge, 1998), p. 35. Translator's Note.]

7. Giordano Bruno, *De gli eroici furori*, in *Dialoghi italiani*, ed. G. Gentile and G. Aquilecchia (Florence, 1957), p. 997. [English version, *The Heroic Frenzies*, trans. Paul Eugene Memmo, Jr (Chapel Hill, 1964), p. 116. Translator's Note.]

8. Bertrando Spaventa, *Rinascimento, Riforma, Controriforma* (Venice, 1928), pp. 229–38.

within the limits of a syncategorematic infinity. But by employing the faculties of the intellect, and above all of the mind, we may achieve a qualitative leap that cancels the imperfection of the unlimited in a simple intuition. And this intuition *'without being preceded or accompanied by discourse* ... comprehends all things; it resembles a vital and smooth mirror, which is, simultaneously, time, light, mirror and all the images that one sees without succession of time or alternation, as if the head were all eye, or as if everywhere the sight embraced in a sole act all things – higher, lower, before and after.'[9] The attainment of actual infinity thus unfolds in an instant; and the only duration proper to it is instantaneous, since it can never be represented as the final term of a potential evolution. Any series that precedes it is irrelevant because it does not condition it in any way.

In his dialogue *On the Infinite, the Universe and Worlds* (*De l'infinito, universo e mondi*), Bruno uses five terms in discussing the paradoxical correspondence between the potential infinity to which our imperfection limits us and the actual infinity that contains it. The divine essence is described as the 'boundless boundary of a boundless thing' (*termino interminato di cosa interminata*). Bruno seeks to reveal the meaning of this perfect definition when he writes: 'A boundary, I say, without bound, because one should differentiate the one infinity from the other.'[10] In what does this difference consist? Bruno continues: 'I say that the universe is entirely infinite because it has neither edge, limit, nor surfaces. But I say that the universe is not all-comprehensive infinity because each of the parts thereof that we can examine is finite and each of the innumerable worlds contained therein is finite. I declare God to be completely infinite because he can be associated with no boundary and his every attribute is one and infinite. And I say that God is all-comprehensive infinity because the whole of him pervades the whole world and every part thereof comprehensively and to infinity. That is unlike the infinity of the universe which is comprehensively in

---

9. Ibid., p. 188.
10. Giordano Bruno, *De l'infinito, universo e mondi*, in *Dialoghi italiani*, ed. G. Gentile and G. Aquilecchia, p. 381. [English version in Dorothea Waley Singer, *Giordano Bruno: His Life and Thought* (New York, 1968), p. 261. Translator's Note.]

the whole but not comprehensively in those parts we can distinguish within the whole . . .'[11]

In his *On Learned Ignorance* (*De docta ignorantia*) 2.1–4, Nicholas of Cusa accounts for this difference. He writes that the unity of the Universe (by virtue of which Bruno calls it a 'whole') is not free from plurality, that is, from matter; and matter, as Aquinas had taught, determines a contraction of infinite form into the finite forms of existence. The result is a certain intrinsic 'defect', an unsuppressible presence in the Universe of a possibility that cannot expand beyond itself. The infinite act cannot originate from the indefinite tension of a perennial potentiality that strives to exceed itself. This would be like asserting that the false infinity, Aristotle's *apeiron*, could generate the true infinite of unfathomable Non-Being, which metaphysically and logically precedes it. It would be tantamount to asserting that desire – the most common embodiment of the false infinite – can attain its most authentic and invisible object, which clearly can be attained only after all desire has been suppressed. Hence, Nicholas of Cusa concluded: 'Although with respect to God's infinite power, which is unlimitable, the universe could have been greater, nevertheless since the possibility of being, or matter, which is not actually extendable unto infinity, opposes it, the universe cannot be greater.'[12]

Still, divine perfection is the 'boundless boundary of a boundless thing', the extreme limit of all unlimitedness; infinite as well, but by virtue of its absolute transcendence with respect to that infinity of which it is the insuperable boundary.

The actual infinite is described by Nicholas of Cusa as the absolute maximum, defined as 'what cannot be larger', that is, as the ultimate configuration of stability, or absolute repose, in which every possibility of successive increases or diminutions vanishes. The true infinite is the authentic resolution of the 'more and less', of the 'large and small' (*mega kai mikron*), by which Plato denoted the potential infinite; and for this reason, all proportion and comparison cease to exist in it.

---

11. Ibid., p. 382. [English version, ibid.]
12. Nicholas of Cusa, *De docta ignorantia* 2.1; Italian translation in *Opere filosofiche* (Turin, 1972), p. 111. [English version, *On Learned Ignorance*, trans. Jasper Hopkins (Minneapolis, 1981), p. 90. Translator's Note.]

Nicholas of Cusa writes: 'Such a maximum as is the absolute exemplar cannot be more or less than any admissible exemplification.'[13]

Bruno's phrase, 'the boundless boundary of a boundless thing', implies meanings that reflect the choice of a perspective in which this idea of perfection – as being contained in the invisible dissolution of all unlimitedness – suggests God's power to limit and circumscribe. Embracing a thesis of Melissus, Nicholas of Cusa wrote that a double infinity can be found everywhere – a 'defining infinity' and a 'definable infinity' – and that the limiting action of the former on the latter engenders the finite entity from its infinite principle.[14] God is also the infinite boundary of all unlimitedness in his role as creator of forms, since (as the Pythagoreans had already intuited) every form is the limit of an infinity.

In fact, Bruno's acceptance of the infinity of the Universe never precludes his respect for the Aristotelian views that an actually infinite magnitude cannot be attained and that the infinites grasped by reason are purely potential in nature.

All the same, we detect a sense of the infinite as authentic liberation in Bruno's Lucretian religiosity and in his vision of an unlimited void filled with countless nuclei charged with creative power. And his tone is undeniably harsh and adamant when he refutes the Aristotelian notion of 'an essentially static nature, changing only within the insuperable limits of theologically and teleologically permanent forms'.[15] Bruno's religious sense of limit and of its symbolic plenitude is probably weaker than that of his Greek model, although we never find him indulging in a trivial celebration of pure and simple infinite extension, of the sublime conceived of as the product of an indefinitely expanded dimension. 'Just as God's greatness in no way consists in its corporal dimension,' he writes, 'so we must not think that the magnitude of his image consists in a larger or smaller mass.'[16]

13. Nicholas of Cusa, *Idiota de sapientia et de mente* 2; Italian translation in *Opere filosofiche*, p. 459. [English version, *The Layman on Wisdom and the Mind*, trans. M. L. Führer (Ottawa, 1989), p. 47. Translator's Note.]
14. Nicholas of Cusa, *Il Principio* 33; Italian translation in *Opere filosofiche*, p. 738.
15. The phrase is that of Werner Jaeger; see Roberto Mondolfo, *L'infinito*, p. 155.
16. Giordano Bruno, *De l'infinito, universo e mondi*, in *Dialoghi italiani*, p. 377.

For Bruno, the infinity of the Universe is the product of the infinity of God, who cannot confine his omnipotence within the limits of a closed space or within a finite multiplicity of substances. As we have seen, this view had already emerged when Christian theologians condemned the theses of Aristotelian and Averroist philosophy, which had inspired admiration for a cosmic order founded on limit, and hence 'horror of the infinite' (*horror infiniti*), which was merely an indication of the fatal imperfection and falseness of every infinite that was sought in created nature.

Yet for Bruno the infinite of God is more discernible as an unfathomable enigma than as an extreme reached by the unlimited development of a potential infinity. If we are able to penetrate this mystery, it is not so much by our mere perception of a boundless world as by our ability to read the deepest meaning of the supernatural signs that are woven into its fabric. As Frances Yates has shown, Copernicus' discoveries essentially gave Bruno a pretext for a new 'hieroglyphic' description of the Creation inspired by Hermetic themes, although they failed to furnish the long-awaited scientific proof of the celebrated infinity of the Universe.

Still, the hieroglyphic writing *par excellence* – which used visible signs of a highly abstract and essential transparence to illustrate the mysteries contained in universal archetypes – could only be realized in the pure forms of mathematics and of geometrical images. For Bruno, mathematics was not an autonomous science from which one derived *a posteriori* a metaphorical aid for grasping metaphysical truths. Rather, from the first it formed a part of contemplative philosophy, a science of formal causes, and the art of *a priori* knowledge of principles – just as it had for Nicholas of Cusa or Raymond Lull.

For example, Nicholas of Cusa had geometrically illustrated the actual infinite in the concept of rectilinearity, or rather, in the ultimate, paradoxical resolution of a curve into a straight line. Indeed, the straight line possesses a 'limit' property like that which allows us to define the absolute substratum of every accident. In a straight line, all 'curvature' in fact vanishes, and the 'more and less' of what can be more or less curved ceases to exist. To use equivalent terms, the indefinite capacity of potential infinity for continual augmentation or diminution is resolved in the only geometrical shape which no longer admits either of the two.

To assert that 'the Absolute is the measure of all things' states the same truth as asserting that 'the straight line is the measure of everything oblique.' In fact, implicit in the phrase 'more and less' is a criterion of measure that postulates the existence of an invisible entity – an entity to which one can refer in saying 'more or less', but which is itself not 'more or less' than something else. A curve is more or less curved according to its greater or lesser conformity to a straight line, that is, to something that lacks any of the characteristics of curvature.

When does a curve finally coincide with a straight line? This result cannot be imagined either as a physically or rationally verifiable event, or as an actually realizable geometric shape, since the straight line is never the ultimate term in an indefinite series of decreasingly curved lines. Yet the straight line remains the ultimate term of reference, of measure and of comparison for an infinity of curved lines.

The same must be said for the squaring of the circle, which can only be intuited by 'eyes' that see beyond the possibilities of the senses and of reason. 'Those who have sought to square the circle,' Nicholas of Cusa wrote, 'have assumed that the circle and square coincide in an equality that clearly cannot be shown in sensible figures. For one cannot in fact construct a material square which equals a given circle. They did not intuit their presumed equality with the eyes of the flesh, but with those of the mind; and they strove to see this equality rationally. But since reason does not permit the coincidence of opposites, they failed in their object. Their intellect should have sought this coincidence in a circle which is equal in all polygons, and even equal to a polygon with a different perimeter. Then they would have achieved their object.'[17]

---

17. Nicholas of Cusa, *Il complemento teologico* 4; in *Opere filosofiche*, p. 617.

Hence, it is with the eyes of the intellect that we must explore the rationally unimaginable instance of the coincidence between a curve and a straight line. This can happen in two situations that are only apparently distinct: in the infinitely large and in the infinitely small. In the first case, we can imagine a circumference which, as the length of its radius is indefinitely increased, tends to merge with any straight line tangential to it. In the second case, we can imagine an arc of circumference infinitely reduced, so that it becomes indistinct from the chord that subtends it. Mathematical perfection consists in this reciprocal commensurability and comparability (*adaequatio*) between what is straight and what is curved. The intellectual intuition to which we must resort then grasps what does not exist as a concrete limit. 'It is necessary for me,' Nicholas of Cusa writes, 'to resort to intellectual vision, which succeeds in seeing how the minimum (if unspecifiable) chord coincides with the minimum arc.'[18] The intellect must consider necessary what it knows cannot be demonstrated in a legible diagram, since 'neither the arc nor the chord (if they are quantities) can simply be actually a minimum, since the continuum is always divisible.'[19]

Giordano Bruno examines the same truths, and places particular emphasis on the opposition between straight and curved, viewed as synonymous with terms that underlie every fundamental polarity of existence. Like Nicholas of Cusa, he emphasizes the coincidence of the minimum arc of circumference with the minimum chord, as well as the existence of an invisible unit of measurement that is common to both straight and oblique.

Far from being interpreted as a mathematical imprecision or as a precocious exploration of the realm of analysis, this last point is merely an illustration of the ultimate metaphysical principles formulated using the only descriptive terms compatible with them at the level of abstraction and simplicity.

In his *Introductory Lectures on Geometry* (*Praelectiones geometricae*), for example, Bruno says that the straight cannot be distinguished from the curved by using only one or two contiguous points.

18. Nicholas of Cusa, *De mathematica perfectione*, in *Opera* (Basel, 1565), p. 1110.
19. Ibid., p. 1121.

Instead, the difference between what is curved and what is not only begins to emerge from differences in the configuration of three contiguous points.

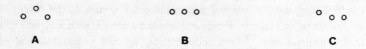

We discover the meaning of this geometric image when we rethink it in terms of an analogy. There exists a geometric minimum of two contiguous points, in which we cannot distinguish the straight from the curved, since they are identical in both cases. By the same token, we may intuit an element antecedent to the variety of forms, a substantial entity that is assimilable to pure quantity, as yet unmarked by the reciprocal distinctions of genus and species. This entity is identical to Plato's indefinite dyad (*aoristos dyas*), that is, to the archetype of every duality; and it can therefore be represented by the geometric 'minimum' of two contiguous points.

By contrast, form begins to reveal itself in constructions made with three points. The number 3 is the return of unity after the unlimited nature of the number 2, and is therefore the archetype of any form considered as a limiting entity, as a barrier against the disintegrating power inherent in *apeiron*.

In his *Idiota triumphans*, Bruno writes that, when we divide an object infinitely, we lose our sense of its quality – that is, of its formal integrity – before we lose our sense of its quantitative substratum, which still remains an essential condition of its existence. Thus, what is determined as form is not yet determined as matter: '*determinatum secundum formam, nondum est terminatum secundum materiam.*'[20] Having passed the cognitive boundary of what is curved, we may still define the geometric 'minimum' that is common to both curve and straight line, and is the principle of all linearity, that is, of everything in the realm of quantities and multiplicities.

---

20. Bruno, *Due dialoghi sconosciuti e due dialoghi noti*, ed. Giovanni Aquilecchia (Rome, 1967), pp. 14–15.

Bruno writes that all the geometric lines that can be drawn according to each genus and species are composed of the straight and the curved, which are the constituent elements of all images, of all figures and of all characters.[21]

Three non-aligned points define a triangle, so that the triangle is the archetype of every plane form, just as the pyramid is the archetype of every spatial form.[22] For Plato, the efficient cause of every formal creation, meaning fire, was composed of minimal elements with a pyramidal shape.

In the *De minimo*, Bruno describes the triangle and the circle as the sources of all figures, because they constitute a bipolarity comparable to form and matter, to act and potentiality, to limit and the limitable (or the unlimited).[23]

In describing the containing power of the circumference with respect to all polygons, of which the triangle is the first and the furthest from the circle, Bruno's statements recall analogous conclusions reached by Raymond Lull in *De quadratura et triangulatura circuli*. In this treatise, Lull sought to solve the impossible problem of the squaring and triangulating of the circle. Using analogous procedures, Lull sought to use 'all' the polygons that can be successively inscribed in a circle, to derive a unique square and a unique triangle, whose areas are presumably identical to the 'last' of the inscribed polygons, that is, with a circle. The result is an incomparable description that geometrically explains the principle of creation and how created forms partake of their ultimate source.

According to Lull, before a circle is 'deformed' by the divisions of a polygon inscribed within it – for example, an inscribed equilateral triangle doesn't divide it into equal parts – it signifies a unique general truth which is diffused and known in a multiplicity of truths as refracted by the soul, a unique and general measure divisible into the multiplicity of partial measures. As the circle is gradually filled with inscribed polygons, this multiplicity spreads to the original void, imposing

---

21. See Giordano Bruno, *Praelectiones geometricae et Ars deformationum*, p. 83.
22. Ibid., p. 31.
23. Ibid., p. 59.

measures and divisions that are defined by particular numerical proportions. Thus, the world of forms is articulated in harmony with the development of its geometric models. But as the number of sides of different inscribed polygons increases indefinitely, we can define a unique 'natural' line of fixed length which, divided by four, generates the side of the desired square: the square that, according to Lull, results in the squaring of the circle.

The 'natural' line thus derived from all polygons restores the primordial and absolute measure on which the 'partial' measure of each is based, and without which it would be nothing, just as numerical proportions could not exist without single units.

The same constructions can be repeated using the triangle instead of the square, since these two figures – the square and the triangle – constitute the principal kinds, the archetypes, and the essential points of reference for every other figure. It is only by their mediation that every polygon can be derived from the measure of the circle that contains it. Lull writes that all the particular measures defined by the inscribed polygons are reduced to three kinds of numbers: the number 3, the number 4, and what he calls the 'circular' number.

'There exists a circle visible to the intellect,' Nicholas of Cusa would write, 'which is equal in all polygons, and also equal to another polygon with a different perimeter.'[24] Lull's method for squaring and triangulating the circle was merely an attempt to display this circle sensibly, as well as to illustrate the infinite equality of the divine essence – which Nicholas of Cusa discussed later – and its reflection in the world of forms through the primordial contraction of the triangle and the square.

---

24. Nicholas of Cusa, *Il complemento teologico* 4; in *Opere filosofiche*, p. 617.

7

Equality

In one of the stories in his *Ficciones*, Borges imagined that a contemporary French writer attempted a remarkable project. Without sacrificing his spontaneous inspiration, he sought to reproduce two entire chapters of Cervantes' *Don Quixote* word for word. This arduous undertaking could be simply declared absurd, since its execution was bound by two contradictory laws: the first was the law of free inspiration, and the second the obligation forbidding any deviation from the original model.

In describing this project, it would not be accurate to speak of an original model and a modern version which, when compared to it, is reduced to a mere imitation. This is not a case of imitation, much less a case of simple copying. The writer's aim of not compromising his creative imagination protects the modern text from any possible loss of originality; and its value is worthy of comparison with the formally identical work that preceded it.

Is all of this possible? In his own words, the French writer replies: 'My task is not difficult. I only need to be immortal to complete it.' The perfect duplication of an original requires immortality.

Here, then, is a paradox that reveals another name of the infinite: equality. True equality is the unattainable limit point which this contemporary writer aims to achieve without passing through the inexhaustible course of successive approximations. 'The ultimate goal of a theological or metaphysical demonstration,' he maintains, 'is no less anterior or common than this novel which I am now developing. The only difference is that philosophers publish in pleasant volumes the intermediary stages of their work, and I have decided to eliminate them.'[1]

In reality, the contemporary writer's project appears manifestly impossible. But if we are to give a plausible explanation of this obvious fact, we must make some not-so-obvious remarks that address the complex paradoxes involved in the idea of infinity.

In fact, Nicholas of Cusa had already faced the same problem in Book 3 of *On Learned Ignorance* (*De docta ignorantia*), where he discusses the hierarchy of universals. He observes that the countless finite creatures of the universe are arranged in an indefinite vertical series that has no final boundary either at the top or at the bottom. Neither the ascending nor the descending process of contractions can reach an absolute maximum or minimum, since God's infinite power cannot be exhausted by things of the creation. This inexhaustibility is transmitted by God to universals, and becomes an indispensable characteristic of every genus and species. The universe does not attain an absolute maximum; and similarly, genera do not attain the ultimate boundaries of the universe, species do not fill up the universalities of the genera, and individuals do not exhaust their species' indefinite power of exemplification. Consequently, we may hazard a preliminary but cautious inference: within the limits sanctioned by their very definition, the universe, genus and species are actual infinites; and by virtue of their proper nature as a synthesizing unity, each of them contains and limits a potential infinity of objects.

Describing the arrangement of species in the order of the universe, Nicholas of Cusa continually appeals to the idea of number. He writes: 'Species are like a numerical series that progresses sequentially and is necessarily finite. *Through them there is order, harmony, and proportion in diversity . . .*' But in what exactly does this diversity consist? From the arguments that follow, one deduces that the necessary diversity of created things – no one thing is ever completely like another – simply verifies a law connected with the problem of continuity.

---

1. Jorge Luis Borges, *Ficciones*; Italian translation, *Finzioni* (Turin, 1971), p. 39. ['Pierre Menard, Author of *Don Quixote*', trans. Anthony Bonner, in Borges, *Ficciones*, ed. Anthony Kerrigan (New York, 1962), p. 49. Translator's Note.]

Nicholas of Cusa expresses it as follows: 'Species are like numbers that come together from two opposite directions, proceeding from a minimum which is maximum and from a maximum to which no minimum is opposed. Hence, there is nothing in the universe which does not enjoy a certain singularity that cannot be found in any other thing, so that no thing excels all others in all respects or excels different things in equal measure, and there can never in any respect be something equal to another. *Even if at one time one thing is less than another and at another time it is greater, it makes this transition with a certain singularity so that it never attains precise equality. Similarly, a square inscribed in a circle passes from being a square which is smaller than the circle to being a square larger than the circle, without ever arriving at its equal.*'[2]

The fact that two distinct things can never be exactly equal is undoubtedly verified. Yet the truth proclaimed by Nicholas of Cusa in the two propositions just cited remains debatable. As the quality of an object changes from one gradation to the next, we cannot immediately assume that it cannot take on all the intermediate values, and that one of these values (or, in fact, an infinite number of them) may be 'singular', i.e., impossible and incompatible with actual existence.

In the nineteenth century, Richard Dedekind formulated a principle of continuity that predicted the formally definable 'existence' of such singularities. And he extended the notion of number from the rational realm to the real one by means of his idea of a 'cut' or section. But he did this at the cost of interminable explanations about the real nature of entities thus defined, and it remained impossible to abandon the notion that irrational numbers were in some way linked to the indefinite.

In Nicholas of Cusa's work, the impossibility of perfect duplication was implicitly associated with an idea of number conceived in the manner of Eudoxus or Euclid. In Euclidean geometry, it is not possible to construct a segment with a length of $\sqrt[3]{2}$ based on a certain unit,

---

2. Nicholas of Cusa, *De docta ignorantia* 3.1; Italian translation in *Opere filosofiche* (Turin, 1972), p. 161; emphasis added. [English version, *On Learned Ignorance*, trans. Jasper Hopkins (Minneapolis, 1981), pp. 127–8. Translator's Note.]

just as it is not possible to construct a segment with a length of π. But if we adopt a model of space governed by the law of continuity, we tend intuitively to admit the existence of such a segment. Thus, we may imagine a cube with a side of 1 enlarged continuously until it became a cube with a side of 2. As its volume increased continuously from 1 to 8, it would assume an intermediate value equal to 2, to which would correspond a side with the length of $\sqrt[3]{2}$.

In Book 5 of Euclid's *Elements*, definitions 4 and 5, which establish when and how ratios of magnitudes may be compared, do not exclude the case of incommensurable magnitudes as a matter of principle. But the case is not explicitly discussed, and the sense of these definitions does not authorize us to extend the meaning of number by an operation analogous to that of Dedekind. In Euclidean geometry, $\sqrt[3]{2}$ was not a real number, and space was always subdivided into finite proportions.

For Nicholas of Cusa, the space of individuals, species and genera has the same structure: such entities are comparable in rigorously defined proportions that exclude the singularity of equality. 'Individuating principles,' he writes, 'cannot come together in one individual in the same harmonious proportion as in another; thus, by itself each thing is one and is perfect in the way it can be. Without doubt, in each species – e.g., the human species – we find that at a given time some individuals are more perfect and more excellent than others in certain respects. Solomon excelled in wisdom, Absalom in beauty, Samson in strength; and those who excelled by virtue of their intellect deserved to be honoured above the others. *Nevertheless, a difference of opinions – in accordance with the difference of religions, sects, and regions – gives rise to different judgments of comparison, so that what is praiseworthy according to one opinion is reprehensible according to another.*'[3]

For Nicholas of Cusa, the absolute equality of any circle with all the polygonal figures through the mediation of the square is geometrically impractical in terms of ordinary, sensory appearances. The squaring of the circle that would define the perfect and infinite Equality can only be conceived as an ideal event, an invisible centre of oscillation of the

3. Ibid., p. 163; emphasis added. [English translation, p. 128.]

perpetual alternation of the 'more or less', of the 'large and small' (Plato's *mega kai mikron*). When we enter the world of forms, this Equality is fatally transformed into duality, and is at best a recognized and definable orientation of this duality that leads back to the primordial unity.

Yet in the geometrical procedures set forth in the *Quadratura circuli* (*Squaring of the Circle*) we see geometrical figures that represent, first, the possibility of straightening a circle and, secondly, an outer circumference made equal in length to a given segment. Both these cases of 'equality' are derived from an implicit principle of continuity. The first case can be illustrated in the absence of 'empty spaces', that is, in the 'completeness' that the intuitive idea of continuity posits as an essential feature of any straight line segment. The second case translates into the evident 'existence' of a point of intersection between two segments. In both examples, the singular case of equality is reflected in something that can be likened conceptually to Dedekind's 'cut'. It figures as an invisible point of mediation between two opposite classes of geometrical shapes: one in which equality is not attained through defect, so to speak; and one in which equality is not achieved through excess. The 'large and small' (*mega kai mikron*) find their centre of equilibrium in a geometrical point whose existence, while intuited, is not rigorously proved. The squaring and triangulating of the circle are assumed even before they are geometrically illustrated, and their feasibility is certainly not the final step in an irrefutable proof. The squaring of the circle has the same obviousness as a point that is intuitively visible on a plane, but also has the same problematic nature. ('Whoever says there is no equal mean between a lesser and a greater quantity,' Nicholas of Cusa writes, 'says there can be no triangle with the same perimeter as a circle. But I assume that the squaring of the circle is possible and consequently assume all the conditions without which it is not possible.'[4])

It is clear that if we remove from such singular cases of equality what might be called their transcendent nature – whatever is simply

---

4. Nicholas of Cusa, *Quadratura circuli* (*Squaring the Circle*), in *Opera* (Basel, 1565), p. 1097.

irreducible to the intermediate configurations into which the comparison of 'large and small' is subdivided – we are faced with a paradoxical game, an antinomy, or an absurdity, from which may spring a duality that is a monstrous counterfeit of Equality.

In fact, if we reconsider the project of Borges' French writer, we cannot help but perceive the ambiguity of the elusive mirage that is engendered by its perverse game of duplicity and pointless repetition. We would seem then to grasp that violating the impossibility of the actual infinite – only an actual infinite would make absolute equality possible – engenders a sort of monstrous caricature of the model that we seek to realize. The lucidity of the project is perverse, in part because it is conducted with meticulously analytical precision. According to the writer's intention, the work can be completed only as the final act of an activity that never moves outside itself. There is not even a conception of any formal totality of the work, which could constitute what might be called a transcendent result. The writer has selected Chapters 9 and 38 of *Don Quixote*, Part 1, and a fragment of Chapter 22.

Are there other examples of such caricaturing duplications, of such counterfeits of perfect Equality? We need only reflect that the ultimate consequence of any antinomy is essentially a form of duplicity, a mirror-image correspondence between two opposing theses, both of them demonstrable and both refutable. Of all the antinomies, those of Kant's transcendental dialectic are the most celebrated. But a complete list would be endless and perhaps pointless, since many antinomies in fact imitate a common model. For example, we could trace back to a single paradigm many different logical and semantic paradoxes that vexed mathematicians after Cantor and the first 'ingenuous' exposition of his theory of sets.

In Chapter 51 of *Don Quixote*, Part 2, one such paradox – analogous to the celebrated Paradox of the Liar – figures as one of the principal riddles proposed in order to challenge Sancho Panza's governorship. The man who formulates the paradox speaks truth and falsehood at the same time; and since the penalty for perjury is death, Sancho in his wisdom decrees that the false 'liar' be cut in two: one part to be released, and the second to be hanged. This division is patently impossible, and in the end it is replaced by an act of clemency. But in any case it emblematically evokes the binary structure of the narration, the continual reference to reciprocal pairs of analogous and balanced events, and the duplication necessitated by the reversal in perspective from Part 1 to Part 2 of *Don Quixote*.

Thus, Borges' writer could not have made a more felicitous choice. But there is a final aspect of the question which deserves brief consideration. Borges hints that the success of the project is favoured by a judgement of contingency about Cervantes' work: while it is true that Cervantes wrote *Don Quixote*, in fact he might not have written it. If we chose to review a hypothetical list of the antinomies, searching for an illustration of *contingency* – of a reality lacking its characteristic necessity and unrepeatable uniqueness – we would find an excellent model for comparison in the perplexities concerning free will that Schopenhauer set forth in *Über die Freiheit des menschlichen Willens* (*On the Freedom of Human Will*), an essay that won the 1839 prize of the Norwegian Society of the Sciences. Schopenhauer explains that when we seek a state of authentic freedom that fosters our human volition, we enter an insoluble infinite regression. A human being can do what he wills, but the question 'Can he will what he wills?', if answered affirmatively, prompts the further question 'Can he also will what he wills to will?', and so forth. Investigating free will requires making the will the object of itself, which triggers an inescapable mechanism that entails the never-ending traversal of a potential infinite.

If we succeeded, absurdly, in reaching a final terminus of this traversal, it would identify the true state of our freedom as absolute indifference, as the absence of any sufficient and determinant reason. And this state of freedom – the 'free will of indifference', *liberum arbitrium indifferentiae* – would, oddly, coincide with total contingency. Between two opposite calls to choose and act, a free man would not in fact know what to do, because he has no clear-cut preference or instinct. By definition we call something contingent when it bears no relation to a cause or reason that lends it the appearance of necessity. A truly free person would find himself condemned to an unresolvable 'duplicity', in the exactly precise meaning of the word, as Dante, for example, managed to express it in *Paradiso* 4.1–3:

Intra due cibi, distanti e moventi
d'un modo, prima si morría di fame,
che liber'uomo l'un recasse ai denti.

(Between two foods, distant and appetizing in equal measure, a free
man would die of hunger before he would bring one of them to his
teeth. [Singleton's translation.])

But what is the source of this illusion of freedom, of this error that
leads us to believe that we are free as human beings? Schopenhauer
identifies the source of this equivocation in what he considers the initial
phase of the act of volition, the stage in which the will is becoming and
has not yet been changed into *resolution*. This stage is *desire*. When a
man simply desires and has not yet decided towards what object he
should direct his action, he thinks that opposite volitions can simultaneously co-exist in his consciousness, and he deceives himself in
thinking that he can extend this duplicity into the next and final phase
of the act of volition. In reality, we can *desire* opposite things at the
same time, but it is possible to *will* only one of them. Hence, there
emerges that necessity that frustrates our search for the 'free will of
indifference' (*liberum arbitrium indifferentiae*) as a state we can actually attain in the world of phenomena. It is no coincidence that the
typical source of false infinity that Leopardi and Hegel identified with
desire is in some way also linked to the antinomies of freedom.

When we desire two objects at the same time, we are creating a
psychological state in which duality can become a potentially paralys-
ing element. If this duality is extended beyond the boundaries of the
desire that engenders it, and invades the sphere of decision and final
resolution, it can prompt a real state of unresolved indecision in the
absurdly free man who faces two antithetical choices. D'Annunzio's
novel *Il piacere* (*Pleasure*) describes the paradoxical state of mind of a
man who over-indulges in the duality envisioned by his desire. If for
no other reason, the book could be read for its description of the
countless sophisms that desire inspires, and their devastating effect on
the hero. Doesn't freedom exist, then? This is clearly not Schopenhauer's thesis. What he proves is merely that freedom cannot be verified
in the ordinary world of phenomena. In this sense, freedom must

simply be relocated: it is 'transcendental' and is essentially a mystery. The consciousness of our original autonomy accompanies all our actions, Schopenhauer recognizes. 'But its true content exceeds the realm of actions and goes much further, since in reality it embraces our being and our very essence, from which (by the impulse of motives) all actions necessarily spring.'

In his *Tractatus* 5.632, Wittgenstein perceived the characteristics of a *limit* in the autonomy of the subject's will. 'The subject does not belong to the world,' he would write, 'but is a limit of the world.'[5] We are obviously dealing with the limit of the infinite regression envisioned by Schopenhauer. And Simone Weil repeated this as well: 'La liberté est une limite.'[6]

This limit point, where necessity and freedom finally coincide in the essential being of the subject, is also a point of connection, of a harmonic relation between the poles of the dilemma. In terms of freedom-necessity, this limit point repeats the function of what Gioberti called 'methexis' – participation, relation, since the indispensability of its existence is 'created' by intermediate terms (*ta metaxy* in Greek). Gioberti had emphasized that the act of creation bore special witness to this connection *par excellence*, the prototype of harmony. And in this infinitesimal movement, which causes multiplicity to spring from unity, and linearity from a point, we may also find a prime model of the act that would reconcile necessity and freedom.

Henri Poincaré and Ferdinand Gonseth perceived a reflection of this model in those moments of spiritual life that reveal authentically creative activity. In *La science et l'hypothèse*, Poincaré suggested that even mathematical invention is capable of displaying this potential and converting it into incontrovertible fact. When a mathematician formulates the postulates of a theory and proposes an axiomatic system, he occupies the precise point in which the *freedom* of invention (for axioms are neither synthetic judgements nor experimental data) is reconciled with the character of rigorous necessity in a chain of success-

---

5. Cf. also Wittgenstein, *Lezioni e conversazioni* (Milan, 1976), pp. 35–6.
6. Simone Weil, *Cahiers*, vol. 1 (Paris, 1970), p. 35. [Weil, *First and Last Notebooks*, trans. Richard Rees (Oxford, 1970), p. 26: 'Liberty is a *limit*.' Translator's Note.]

ive deductions. But what can this fact mean, Gonseth comments, except that the spirit has *freely* succeeded in making itself *determined* by things? '*Evidence*,' he writes, 'is the point at which the spirit binds its freedom to the determinism of things.'[7] If we consider the evidence of a proof and the clear connection between a premise and its conclusion at every step, or between the free positing of the hypothesis and the rigorous necessity of deduction in the thesis, then this evidence appears to be the sign of a possible connection between the phenomenal and the subjective, between the determinism of the world and the autonomy of the spirit as the prerogative of the moral realm.

The problem of the 'free will of indifference' was later touched upon by Saul Kripke in the essay 'Identity and Necessity'.[8] He poses the question: can the *same* man be contemplated in *different* possible worlds, that is, in different counterfactual conditions? Can we ask the question 'Could a certain person have behaved in a way different from previous behaviour, and still have remained the same person?' Kripke answers this question in the affirmative by distinguishing between *rigid designators* and *non-rigid designators*. The former denote an object that remains identical in all possible worlds: e.g., the square root of 25. The latter refer to an object that can vary according to different possible circumstances. In the real world, 'the orator who unmasked Catiline' refers to Cicero, but in a counterfactual situation it could refer to a different person. Kripke is inclined to regard proper names like Cicero, Franklin and Nixon as rigid designators, and to regard eventual defining circumstances (e.g., 'the 37th President of the United States') as *a posteriori* truths susceptible of denoting the object in 'non-rigid' fashion. According to Kripke, Nixon might *not* have done everything that he did, with one exception: not to be Nixon. The question 'Could Nixon have been a person different from what he actually was?' is ambiguous because it suggests a literal sense in addition to the seemingly acceptable metaphorical sense that 'Nixon could have been a different *type* of person.' The latter seems to be a plausible question, while the literal sense is self-contradictory. But the

---

7. Ferdinand Gonseth, *Déterminisme et libre arbitre* (Neuchâtel, 1947), p. 186.
8. In *Identity and Individuation*, ed. Milton K. Munitz (New York, 1971).

problem is precisely this: to what extent does the metaphorical sense protect us from an intrusion of the literal sense?

His allusion to free will leads us to think that Kripke does not actually intend to give an answer, or at least that he is most concerned with demonstrating the logical consistency of a language in which modal expressions refer to individuals who remain invariable in different possible worlds. He implies this when he develops formal systems which include modality and quantification, and in which theorems of completeness can be formulated, as in his article 'A Completeness Theorem in Modal Logic'.[9] Still, there remains an irrefutable datum. In support of his models, Kripke defends Aristotle's essentialism, although he reconciles it with empiricism in a singular way, and in some cases derives it from *a posteriori* truths. (If by an empirical judgement, I recognize that *this* desk is made of wood, he maintains, being made of wood is an *essential* property of *this* desk; and it could not have been made of ice, for example, without changing into another desk.)

Schopenhauer's position is also at heart essentialist. In his essay on the freedom of the will, he concludes by locating the principle of freedom in the innermost quality of the subject, and thus making it an aspect of its *essentia*. Not unlike Kant, he saw a reconciliation between necessity and freedom in the encounter between the inevitably empirical action and the transcendental nature of the subject. Still, if realized, this 'coincidence of opposites' (*coincidentia oppositorum*) would lead to unimaginable conditions. Man is intrinsically free, because after all 'the door was never closed', as a Sufi master expressly put it, or as we read in the Book of Revelations 3.8 ('I have set before you an open door'), or as Kafka suggested in the conclusion of *The Trial*. But we are in fact dealing with a 'truth without form' and with a dangerous void, and any attempt to transpose them into tangible circumstances risks changing them into a false ethical miracle, into pure nonsense. Nietzsche did not fail to attack the vestige of empirical cognition that Schopenhauer had saved for free will. In *Human, All Too Human* 1.39, he accused Schopenhauer of chopping logic by arguing that our freedom involves our very being because we impute our actions to

---

9. *Journal of Symbolic Logic* 24 (1959): 1–14.

ourselves and cannot escape our sense of responsibility. It is true that some of our actions may cause us to feel 'discomfort' or a 'sense of guilt', Nietzsche maintains, but it is absurd to infer from the *fact* of our discomfort 'the reasonableness and rational *admissibility* of this discomfort', and to interpret it as a sure sign of our *essentia*. 'It is because man *considers* himself free, not because he is free,' he concludes, 'that he feels repentance and remorse.'

But essence, Kripke insisted, is a necessary substratum of discourse. Even if we call it a fiction and follow Quine in reducing the ontological implications of language as far as possible, it is still difficult to ignore the unifying and conciliatory function of the elements that compose a unit of discourse. For Kripke, when we introduce necessity and possibility, and enter into modal logic, it becomes difficult to sustain the thesis that merely envisioning counterfactual situations or different possible worlds undermines the integrity of an essence. When we say that in different circumstances Cicero could have done things differently, we are speaking about a unique and indivisible individual, precisely about *him*, that is, about Cicero, invoking the identity principle that A = A. The unlimited explosion of foreseeable possibility, the uncontrolled imagination of counterfactual conditions, that risky 'sense of possibility' described by Musil in one of the first chapters of *The Man Without Qualities* – all these are necessarily counterbalanced by a unit of reference based on an immutable central subject, whose only role is to *limit* this potential infinity in some way. There is still a question of choice here. Either we accept this limit, which is also a sort of invisible actual infinity, since it is the analogue of a substratum of all accidents; or else we reject it as a fiction, censuring it for this *negative* quality as a non-representable object. Kripke took the first route, Quine the second. But it is also clear that our censure should not exclude anything that can make a fiction the pivot and source of all factuality.

Quine was also substantially sceptical about the use of equality and about the consistency of a quantified modal logic, since he found that this implied an acceptance of Aristotelian essentialism. He explained that the modal contexts 'It is necessary that . . .' or 'It is possible that . . .' lead to ambiguities and contradictions when combined with substitutive applications of equality: as a result, he called these contexts *referentially opaque*. For example, if we say that '9 is *necessarily* greater than 3', our statement is true. But then if we discover that 'the number of planets in the solar system is 9' and if we claim that 'the number of planets in the solar system is *necessarily* greater than 7', we lapse into error, since we transform an *a posteriori* truth, which is clearly a scientific discovery, into a necessary and apodeictic truth. Quine did not even think that necessity could be reconciled with what is *a posteriori* and with the empirical truth, which he associated with contingency. For the most part, his examples of *referential opacity* operate by substituting in a modal context (or in a context featuring expressions like 'he believes that . . .' or 'he knows that . . .') the terms of an equality based on empirical observation. And an analogous device has the effect of rendering useless the remedy suggested by the so-called 'individual concepts' of Carnap and Church.[10] In this way, referential opacity means in part recognizing the descriptive complexity of some names of objects, and the fatal inclusion, in some of them, of empirical circumstances which corrupt the naked essence that the name denotes. By using the 'number of the planets' to denote the number 9, we would combine *a posteriori* contents, derived from astronomical observation, with certain qualities that we must also recognize in the number 9 in order to be able to ascribe a character of necessity to arithmetical propositions like '9 is greater than 7'.

But Quine's arguments lose some of their cogency if we accept the impartial use of essentialism proposed by Kripke. The property of an object, he observed, may be considered *essential* even when it has been revealed by an empirical observation. Essentialism and necessity are not at odds with what is *a posteriori*; and this latter term is not synonymous with contingency. Thus, it is a plausible thesis that some statements of identity, even if recognized empirically, are necessary, provided they are true. By thus expanding the domain assigned to *essentia*, we are able to counteract the referential opacity of modal contexts and the ambiguities of equality, and able to confirm the logical reproducibility of ever greater parts of ordinary languages.

---

10. See W. V. Quine, *From a Logical Point of View* (Cambridge, Mass., 1961), Chapter 8; Italian translation, *Il problema del significato* (Rome, 1966), pp. 129–48.

8

Descartes

In replying to objections that the theologian Caterus had voiced against certain ideas expounded in his *Meditations*, Descartes made clear *inter alia* what he meant when he used the word 'infinite'.

The distinction that he makes here between infinite and indefinite reflects the traditional opposition between actual infinity and potential infinity, but the terms of the problem appear to have shifted slightly. The word 'indefinite' still evokes the fatal imperfection of the terrestrial object that seems to lack an internal limit. But 'indefinite' now refers typically to something which, while limitless in some respects, is still subject to limitation, since it cannot escape the boundaries created by its own particular existence, by its being what it is and not something else.

'I distinguish here,' he writes, 'between the indefinite and the infinite. Strictly speaking, I designate only that thing to be infinite in which no limits of any kind are found. In this sense God alone is infinite. However, there are things in which I discern no limit, but only in a certain respect (such as the extension of imaginary space, a series of numbers, the divisibility of the parts of a quantity, and the like). These I call indefinite and not infinite, since such things do not lack a limit in every respect.'[1]

From Descartes' other more explicit statements, it is easy to understand that this simple distinction between infinite and indefinite contains a certain element of novelty.

---

1. René Descartes, *Œuvres et lettres* (Paris, 1953), p. 352. [English version in Descartes, *Philosophical Essays and Correspondence*, ed. Roger Ariew (Indianapolis, 2000), p. 155. Translator's Note.]

In fact, from the Cartesian perspective the imperfection denoted by the term 'indefinite' consists more in a residual presence of limit than in a continual opening towards overcoming it. It was this same continual opening which Boethius had rejected as a 'monster of malice', and which Aristotle had associated with non-being and privation. Instead, in the third of his *Meditations* Descartes perceives in it the unequivocal sign of a divine imprint. 'To be sure,' he writes, 'it is not astonishing that in creating me, God should have endowed me with this idea, so that it would be like the mark of the craftsman impressed upon his work, although this mark need not be something distinct from the work itself. But the mere fact that God created me makes it highly plausible that I have somehow been made in his image and likeness, and that I perceive this likeness, in which the idea of God is contained, by means of the same faculty by which I perceive myself. That is, when I turn the mind's eye towards myself, I understand not only that I am something incomplete and dependent upon another, something aspiring indefinitely for greater and greater or better things, but also that the being on whom I depend has in himself all those greater things – not merely indefinitely and potentially, but infinitely and actually, and thus that he is God.'[2]

From this perspective, the mind's opening towards the unlimited – which is manifested in *desire*, the source of certain noted difficulties of the false infinite – is viewed as God's reflection in man's imperfection. In the third of the *Meditations*, we read: 'For how would I understand that I doubt and desire, that is, that I lack something and that I am not wholly perfect, unless there were some idea in me of a more perfect being, by comparison with which I might recognize my defects?'[3]

Does there, then, exist in man a positive idea of the infinite, which makes limit appear a defect, and desire and doubt symptoms of our desire for liberation? If so, evil and antinomy are not in themselves a reverse symbol of God's perfection, nor does limit appear as a condition of Platonic form (*eidos*), of the divine gaze. The devastating and revelatory power of antinomy is replaced by a positive intuition of the

2. Ibid., p. 300. [English version, pp. 121–2.]
3. Ibid., p. 294. [English version, p. 118.]

totality animated and nourished by desire, a potential corrective for a perennially defective creature. On the other hand, there is no longer any trace of that relative perfection of the limited creature in which Nicholas of Cusa had discerned the image of a 'finite infinity', of a formal plenitude blessed by the divine eye. God in fact no longer loves limit.

In a letter that Descartes wrote to Pierre Chanut on 6 June 1647, he begins to describe the world as a boundless network of functional relations with no orientation towards the absolute. The declaration that God alone is the final cause (*'omnia propter ipsum [Deum] facta sunt'*) is used as irrefutable proof that no creature, not even man, can be considered the definitive goal of any teleological order. Each thing is susceptible of increase or decrease; and rather than closing within itself, in its form or limit, it is at once measured against something that is *other*. Pascal would later develop fully this tendency of the spirit, venturing to refer each thing, and principally man, to this infinite totality of correlated events. In this view, any object whatsoever thus becomes unknowable: for to apprehend it would require perceiving all its mediate and immediate causes, all the bonds that bind it to more distant and undiscoverable facts. As the Pythagoreans had said, what is unlimited is unknowable.

Matter, separated from the spirit, now begins to disintegrate. In the sixth of the *Meditations*, Descartes declares its divisibility the element that distinguishes it from spirit, which is indivisible. This is an obvious and evident symptom that the breakdown of the equation between spirit and world translates into form's loss of integrity, and hence into the prevalence of the unlimited over limit. The reduction of bodies to mere extension offers a promising anticipation of the formal disintegration later effected by spatial antinomies.

Under these assumptions, the indubitable truth that the world is only given to us as an idea, as a subjective representation, takes the particular form of activating the most traditional rules of metaphysics, which everywhere reveal appearances that deceive. Descartes was clearly not the first to encounter this truth, but it is now revealed with its power to empty and demolish, since appearance in fact loses what had earlier pervaded and sustained it. The illusory trick of Nature leaves the stage not because it had attained the liberation that was proclaimed as its ultimate finality, but simply because the gods were eclipsed and the spirit separated from what should have helped reveal it.

Not by chance, Schopenhauer in his *Parerga und Paralipomena* classified Descartes as the first philosopher who succeeded in formulating a clear distinction between *real* and *ideal*. When Schopenhauer summarizes the history of philosophy, he begins with Descartes and ends with a description of Kantian phenomena, according to his aim of depicting the evolution of a unified perspective. This perspective could also be characterized in the following way. Beginning with Descartes, form is emptied of its peculiar content – that is, of its essence and Platonic form (*eidos*) – and thereby ceases to be a revealed spiritual existence, an exterior expression of divine light. Florenskij observed that the antithesis of the Greek *idea* (conceived as a divine gaze, as a ray of the 'source of all images') is represented by the 'mask' or 'larva', a pseudo-real shell that popular wisdom always associates with evil and impurity. The transition from the 'idea' or 'divine gaze' to the 'mask' can then be translated philosophically into the transition from the reality of phenomena, conceived in a popular or Platonic sense, to the Kantian, positivistic and illusory reality of phenomena: two different phases of existence and thought, neither of which lacks its own coherent object of research. 'When the face becomes a mask,' Florensky writes, 'as Kantians we can no longer know the *noumenon*, and as positivists we have no basis for asserting its existence.'[4]

Subsequently, Kant reduced even the primary qualities of every intuited object – such as extension, form and movement – to universal schemes of subjective apprehension. Paradoxically, these very schemes, which on the one hand guaranteed the objectivity of knowledge, were also the principal distorting lens by which one apprehended an object that was by now definitely unknown. Later, Schopenhauer would describe the illusory nature of space and time, precisely by using their being as a basis for mental laws of sensible intuition, that is, for subjective modes of organizing experience. If space and time do not

---

4. Pavel Aleksandrovich Florenskij, *Le porte regali, saggio sull'icona* (Milan, 1977), p. 49.

belong to the thing in itself, but interpose themselves between it and us, it means that in fact their essence is an illusion, since by clothing the reality of determinate forms they also irremediably conceal it from our gaze. The transformation of the object into something quite similar to Zeno's paradoxes, laden with the dangerous ambiguity of a fragmenting antinomy, is tersely evoked by Schopenhauer in his *Parerga und Paralipomena*: 'In a certain sense, the past is *not* past, and everything, which has ever really and truly existed, must at bottom still exist, since indeed time is only like a stage waterfall that appears to flow downwards, whereas, being a mere wheel, it does not move from its place.' And further: 'Long ago, in my chief work, I compared space analogously to a glass cut with many facets which enables us to see in countless reproductions that which exists singly.'[5]

Subsequently, the search for a conclusive truth surpassing the limits of ordinary spatial and temporal intuition evolved in the realm of mathematics. Weierstrass, Dedekind, Cantor and Russell would emphasize the reliability of abstract mental mechanisms that operate outside of space and time. For example, the difference between sensory learning and true knowledge would be adopted by Russell to justify a mathematical definition of continuity. To be acceptable, such a definition must admit an analytical penetration of the real that exceeds the finite threshold of our concrete experience, thus developing and finding its proper support in the pure realm of logic and in accord (or at least not in disaccord) with the facts.[6] The breakdown of the continuum into points joined by invisible relations had ultimately to be realized outside our habitual knowledge; and some of the traditional antinomies found an apparent solution beyond our common sense, which itself would have had to renounce its age-old certainties to conform to new rules and new paradigms.

Nevertheless, the new paradigms would themselves submit to good

5. Schopenhauer, *Parerga und Paralipomena*; Italian translation (Turin, 1963), pp. 129–30. ['Fragments for the History of Philosophy', English version in Schopenhauer, *Parerga and Paralipomena*, trans. E. F. J. Payne, 2 vols. (Oxford, 1974), 1:85. Translator's Note.]

6. Bertrand Russell, *Our Knowledge of the External World*; Italian translation, *La conoscenza del mondo esterno* (Milan, 1975), pp. 140–41.

sense, as Wittgenstein demonstrated. The decisive justification of theorems and proofs arrives at a term that is never truth and absolute evidence, but a reflection of habits acquired by common sense formed by learning and acting.[7] Even before this, Poincaré would observe that Russell's theses were unable to eliminate Kant's apriorism, and after Cantor the intuition of time would still play an important, if not decisive, role.

Descartes was also among the first to intuit the infinite from viewpoints that were destined to develop an eminent role in later history. In a letter to Mersenne of 1630, he refuted a rather wordy argument that proved that infinite sets do not exist. Mersenne had cited the simple fact that an infinite line must contain both infinite feet and infinite fathoms. Since a fathom is six times longer than a foot, the infinite set of fathoms would absurdly have to contain the infinite set of feet as a subset, while both sets would coincide with the infinite line. One could only conclude the following: the infinite line cannot exist: for if it existed, it would have to coincide with each of two infinite sets, one of which is larger than the other.

Descartes calmly accepted the paradox, but denied that one could draw from it the conclusions that Mersenne took for granted. Instead, the paradox revealed a predictable feature of every set that is represented as infinite. The ratio between a fathom and a foot is a *finite* number, which makes Mersenne's observation incompatible with what applies to the infinite and its laws. The norms governing the comparability of infinite sets must completely transcend every finite proportion, such as the one involving the ratio between a foot and a fathom.

In reality, Descartes' objection was profound and grasped one of the knots of the problems. This is shown by the fact that two centuries later Cauchy was to take an example not unlike Mersenne's to prove the non-existence of actually infinite sets. Indeed, Cauchy argued in the following way. If we assume the entire series of whole numbers as given, we can also form a series of the squares of these numbers, so that for each integer n there corresponds a square n^2. The nature of

---

7. Wittgenstein, *On Certainty* (*Della certezza*) (Turin, 1978), p. 35.

this correspondence means that the two series contain the same number of elements, but it is evidence that the set of the squares n^2 is only a part of the 'larger' set formed by all the whole numbers n.

This paradoxical example, which Cauchy attributes to Galileo, was supposed to prove that whole numbers could not be conceived as an actual totality, that is, as an actually infinite set. But Descartes' objection still had good reasons for prevailing. Instead of denying the paradox, Descartes made it an inevitable consequence of the infinite: the infinite is paradoxical precisely because the laws of finite comparability do not fit it.

In fact, Descartes would not have defended the existence of the actual infinite, and he certainly would not have done so using Cauchy's arguments. The formal and limiting element that an actual infinite must possess cannot coincide with the analogous formal content of sets and finite numbers: indeed, it differs from them precisely because it is 'infinite' and not 'finite'. By contrast, Cauchy's thesis implied that if the actual is to exist, it must be finite. In essence, he had thus repeated the simple truth that we cannot count *all* the whole numbers one by one to the point of arriving at a final term, revealing in this way that he felt compelled to conceive the infinity thus reached, in terms of a finite totality – as if, instead of counting all the numbers, he had only counted to 10 or 100.[8]

In the end, Richard Dedekind sought to overturn Cauchy's thesis, using this paradox not as proving the infinite's non-existence, but as essential to its very definition. For Dedekind, infinite sets became those sets that can be placed in one-to-one correspondence with one of their parts, precisely in the way that integers can be placed in one-to-one correspondence with their squares ($n \leftrightarrow n^2$) or with even numbers ($n \leftrightarrow 2n$).

Ultimately, Dedekind's idea proved unexceptional. Among its merits, one could count the definitive refutation of the idea that one must recognize the nature of a 'maximal' entity in any mathematical object that reproduced the infinite. In reality, this possible misunderstanding had already been reported by Kant, in his note on the thesis

8. Louis Couturat, *De l'infini mathématique* (Paris, 1973), pp. 445–6.

of the first antinomy. In that passage, he explains that the concept of infinite does not coincide with that of maximum, but can be conceived only as a relation with respect to a unit *chosen at will*, in the sense that it is greater than any given number of such units. Thus, the infinite does not become larger or smaller in proportion to a chosen unit; rather, being infinite, it remains unvaried despite variations in the unit. Its essence consists rather in the capacity to comprise and contain any quantity that can be attained by the mere procedure of counting. Hegel was to repeat that in such a case we are dealing with a ratio and with a quality, not with a quantity; one cannot understand the infinite by specifying how 'large' it is.

9

Leibniz

Descartes also anticipated the comprehension and assimilation of a second aspect of the infinite which, beginning with Leibniz, was called the 'principle of continuity'.

In a letter to Desargues, Descartes approved his idea that a group of parallel straight lines could be considered a variant of a system of straight lines converging at a single point.[1] If the lines are parallel, we may doubtless attribute to them a common property, which could be unambiguously defined as their 'direction'. But this 'direction' can also be called a 'point at infinity' when we consider the perspectival disappearance of the parallel lines as they recede indefinitely against a real or imaginary background. In fact, the idea of calling a direction a 'point' came to Desargues through his studies of perspective. For a visual ray that follows the common direction of straight lines tends to focus on an endpoint 'at infinity', which seems to be their ideal intersection point.

As Erwin Panofsky once observed, Renaissance perspective played a decisive role in the emergence of a new idea of infinity. Borrowing from Cassirer, Panofsky regarded perspective as one of those 'symbolic forms' by means of which 'spiritual meaning is attached to a concrete, material sign and intrinsically given to this sign'.[2] The origins of this shift had deep roots. The world of the Greeks was essentially discontinuous, and it unfolded through the inexorable antagonism between

---

1. See Louis Couturat, *De l'infini mathématique* (Paris, 1973), p. 264.
2. Erwin Panofsky, *Die Perspektive als symbolische Form*; Italian translation, *La prospettiva come forma simbolica*, p. 47. [English version, *Perspective as Symbolic Form*, p. 41. Translator's Note.]

the formlessness and passive resistance of space and the delineations of form. By contrast, Neoplatonic metaphysics, Byzantine art and Romanesque painting and sculpture revealed a perception of space as a continuous 'quantum', as a homogeneous fluid in increasing harmony with the shapes that emerged from it. The 'infinite' character of this space could not have been unknown to the Greeks, but the importance of this infinity for figurative representation lay entirely in the negative and 'material' nature of *apeiron*.

Plato's space (*chora*) was synonymous with indetermination and substantiality. It was a 'receptacle' of forms, a primary condition of their growth and diversification, and its characteristics were those of infinity. Giorgio de Santillana advances the hypothesis that in archaic thought the world's ordered structure found its primordial form in the cyclical nature of time, and that space itself, by its indeterminate nature, was its antagonist, the principle of absurdity and incoherence. In Plato, he observes, space was identified with Non-Being. The first breakdown of cosmic unity bound by cycles did not occur with Copernicus or Kepler: for 'they both recoiled from unboundedness.'[3] Even Galileo continued to be dominated by cycles and circularity. As de Santillana writes, 'The Untuning of the World, the dissolution of the Cosmos, was to come only with Descartes.'[4] Space thus became the most propitious territory for exemplifying opposite conceptions of the world. Where, for a Greek, could it be less natural to locate the actual infinite (*energeia apeiron*) than in space? Yet it was in space that Ambrogio and Pietro Lorenzetti, Jan van Eyck and Dirck Bouts placed a visible image of the actual infinite as the vanishing point of the lines of depth. The rationalization of space achieved by Descartes, Panofsky writes, finds an ideal premise in this infinite extension into which *positive being* has burst, the affirmation of existence culminating in the *visibility* of a *point* in which the entire infinity of visual space appears enclosed and unified.

In what sense can we assert that points of convergence at infinity are a special application of the principle of continuity? One of the ways in

---

3. Giorgio de Santillana and Hertha von Dechend, *Hamlet's Mill*, p. 342.
4. Ibid., p. 343.

which Leibniz successively formulated this principle is the following. If the difference between two cases or figures can be reduced below a level that is effectively assignable in concrete data, then it is necessary that this difference *can be reduced below any assignable quantity* even in figures which cannot exist *in concreto* but which can only be imagined and obtained as the result of continual variation.

Thus, if we imagine two non-parallel lines in a plane, they will certainly intersect at a point. But if we continuously vary their directions until they become parallel, the point of intersection will indefinitely recede in the plane until it completely vanishes in the extreme case or limit of parallelism. What the lines have in common in all the intermediate configurations must in some way also be present in the final stage of the variation, which is represented by parallelism. The order of data must be communicated in an analogous order, one that is recognizable in the unreachable point towards which they are oriented – a point which in itself is invisible, but which is indirectly revealed by the *uniqueness* of the direction that defines the parallelism.

Points of convergence at infinity can also be justified by their 'visibility' as ordinary geometrical configurations, often by virtue of simple transformations through projection. For example, if we fix some essential points of reference on two lines, the point at infinity on one of them can be made to correspond to a point 'at a finite distance' on the other, and a continuous variation towards infinity on the first can correspond to a finite extension of the second.

Such apparently irrefutable deductions – concerning the 'existence' of the infinite and its attainability through continuous movements – were the result of an idea of the infinite nearly the opposite of that which had inspired Aristotle. The extent of this shift can be best expressed by contrasting two lapidary statements that encapsulate all the possible expressions of these two irreconcilable visions of the world.

The first is taken from Aristotle's *Generation of Animals* 1.1.715b15, which reads: 'Nature avoids what is infinite, because the infinite lacks completion and finality, whereas this is what Nature always seeks.'

The second statement is found in Leibniz's letter of 1693 to the

Marquis de L'Hôpital, and affirms the opposite truth: 'The perfection of the analysis of transcendents and of geometry that treats of infinites will doubtless prove most important because of its possible applications to the operations of nature, which involves the infinite in all it does.' In fact, as Leibniz remarks elsewhere, 'the nature of the infinite Author ordinarily occurs in the operations of nature.'[5]

His wonder in contemplating the infinitely small moved Leibniz to write revealing words which allow us to make inferences about his mathematical innovations. In his *Monadology* 67–8, we read: 'Each portion of matter may be conceived as being like a garden full of plants and like a pond full of fish. But each branch of every plant, each member of every animal . . . is also some such garden or pond.' And he further writes: 'And though the earth and air which are between the plants of the garden, or the water which is between the fish of the pond, be neither plant nor fish; yet they also contain plants and fishes, but mostly so minute as to be imperceptible to us.'[6]

'By envisioning the existence of different orders of infinity,' Léon Brunschvicg observed, 'Leibniz's mathematics made it possible to imagine the real infinite as a simple degree, and to superimpose on it an infinity of possible infinities . . . Divine wisdom, Leibniz explained, goes beyond the finite combinations found in reality; it makes an infinity of infinities, that is an infinity of possible sequences of the Universe, each one of which contains an infinity of creatures.'[7] The mechanisms transforming reality and realizing the possible act within the subtlest interstices of this arrangement. They are grasped, as it were, as directing a nascent state that encloses a maximum of unlimited potentiality joined to a maximum concentration of entelechy, or delimiting essence. Thus, science sets out to seek the characteristics of

5. Leibniz, 'Considérations sur la différence qu'il y a entre l'analyse ordinaire et le nouveau calcul des transcendantes', *Journal des Sçavants* (1694); cited in René Guénon, *Les principes du calcul infinitésimal* (Paris, 1946), p. 45.

6. Leibniz, *Monadology*; Italian translation, *Monadologia* (Bari, 1937), p. 143. [English version, *The Monadology*, trans. Robert Latta (Oxford, 1898), p. 256. Translator's Note.]

7. Léon Brunschvicg, 'Spinoza et ses contemporains', *Revue de Métaphysique et de Morale* 14 (1906): 35–82, at p. 58.

action and change 'in the momentary element, in the force tending towards change', which for Leibniz is the very principle of reality and the profoundest source of its dynamic functioning.

Such infinitesimal forces, more susceptible of an ideal than a real existence, through which concrete form springs from the possible, were simply the differentials dx.[8] Mathematics was the image most like the infinite complexity of the machine of nature.

The very definition of differential, which Leibniz introduced in a purely geometrical manner (rather than in terms of the concept of function), evoked a kind of primitive force, an invisible essence from which one could deduce the effects of the ordinary appearance of phenomena. It is worthwhile to briefly describe Leibniz's procedure.[9]

The reasoning can be developed with reference to a curve C drawn on a Cartesian plane with two coordinate axes x and y. If dx and dy represent two corresponding finite variants of x and y in the passage from point P (with coordinates x and y) to point Q (with coordinates x + dx and y + dy) on curve C, and Δx is a fixed segment, we may define a quantity Δy by means of the proportion

(1)
$$\frac{\Delta y}{\Delta x} = \frac{dy}{dx},$$

where the ratio dy : dx, in terms of familiar trigonometric ratios, denotes the tangent of the angle formed by the straight line c, passing through P and Q, and axis x.

If dx now diminishes arbitrarily until it approaches zero, the ratio dy : dx becomes the ratio of two very small quantities. If dx reaches the point of vanishing definitively, then line c, passing through an infinite number of intermediate passages, finally becomes identical with line t, which is the tangent to curve C at point P. As Leibniz explains, this means that the quantity Δy can be defined even when dx = 0, since in that case the ratio Δy : Δx must coincide with the tangent of the

---

8. Ibid., p. 52.
9. See H. J. M. Bos, 'Differentials, Higher-Order Differentials, and the Derivative in the Leibnizian Calculus', *Archive for History of Exact Sciences* 14 (1974): 1–90.

angle formed by line t and axis x. That is, $y : \Delta x$ always remains a finite ratio, even in the extreme case in which $\Delta x = 0$, so that $\Delta y : \Delta x = y : d$ (where d is the distance between the intersection of line t and axis x and the projection of point P on the same axis).

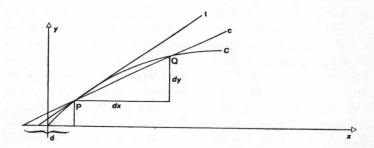

Hence, for $dx = 0$, the ratio $dy : dx$ can still be interpreted, by virtue of proportion (1), as a ratio between finite quantities, and therefore still means *something*. But this *something*, which cannot be expressed by assigning dx a value that is *rigorously* null (which would produce an absurdity and make the problem completely indeterminate), must preferably be associated with a concept that alludes to 'zero' while preserving intact whatever remains visible in a regular macroscopic situation. This concept is the concept of the *infinitesimal*. In this case, we shall say the dx and dy are two *infinitesimals*, and it will become natural to think of the infinitesimal as a *quantity smaller than any assigned finite quantity*.

If we wanted now to name the mechanism that has made it possible to attain the infinitesimal, we would be forced to invoke the principle of continuity. By virtue of the visibility 'in concreto' of the final solution of a variation (the tangent line), this very principle renders significant the ratio $dy : dx$ even in the extreme case where $dx = 0$. The conceptual ambiguity that was always perceived in the definition of infinitesimal lies then entirely in the application of this principle, that is, in the supposition that a certain configuration is posited as the ultimate case, effectively reached, of an infinite series of intermediate cases.

In no way did Leibniz conceive of limit as Weierstrass would have understood it, namely, as a magnitude or figure which it is possible to approach indefinitely without ever reaching it. In Weierstrass's mathematics, the value of the ratio dy : dx became a limit that was situated beyond the series of values tending towards it, and that was obtained by the progressive approach of line c to the tangent t. With this notion of limit, in perfect coherence with the objections that Aristotle himself could have made to Leibniz's theses, it becomes possible to reformulate rigorously the definitions and theorems inspired by the principle of continuity.

Instead, Leibniz conceived of his results as 'extrapolations to the actually infinite of concepts of the calculus of finite sequences'.[10] Posited as the final term of an unlimited process, these results were an apparent demonstration of the actual infinite. From this perspective, a curve could be considered as a polygonal segment with an infinite number of sides. (This was a recurrent idea in the differential calculus.) By the same token, a 'discrete' series of numbers could be extended to the 'continuous' case of an infinite sum of differentials, that is, to the integral conceived as an effective infinite actuality, the tangible result of an evident application of the principle of continuity. Conceptually, this meant a return to the theses that Antiphon had advanced concerning the squaring of the circle: Aristotle's refutation could have been applied with equal efficacy to some of Leibniz's arguments.

In the absence of a more rigorous description of the findings of Analysis using the idea of 'passage to the limit', Leibniz continued to speak of infinitesimals dx as 'fictions' useful in the art of mathematical invention, as imaginary entities that did not necessarily correspond to things actually existing outside the mind that conceived them. Instead of infinitesimals, he wrote, one could have used expressions like 'as small as is necessary so that the error is smaller than any given error'. In other words, the actual infinite, which arose in mathematical procedures involving the infinitesimal, could be replaced by the potential infinite of the proofs by exhaustion of Eudoxus and Archimedes. Leibniz himself was conscious of certain anomalies and ambiguities connected with the use of infinitely small increments. 'Such increments,'

---

10. See H. J. M. Bos, 'Differentials', p. 13.

he wrote, 'cannot be shown by any construction. I am in fact in agreement with Euclid (Definition 5 of Book 5) that homogeneous quantities are comparable only when one of them can become larger than another if multiplied by a number, meaning a finite number. I assert that entities whose difference does not fit this type of quantities are equal ... This is precisely what I refer to when I say that the difference is smaller than any given quantity.'[11]

Thus, the techniques of the new calculus obliged mathematicians to 'neglect' infinitesimals with respect to ordinary finite quantities. This rule became one of the first axioms stated in the first book on infinitesimal calculus, written in the 1690s by Leibniz's pupil, the Marquis de L'Hôpital: two quantities that differ by an infinitesimal are equal. It is clear, nevertheless, that two quantities which differ by another quantity, however small, can only be equal at the price of a logical contradiction or an approximation; and it is certain that infinitesimal calculus was *never* a method of approximation.

Another feature of calculus – its failure to obey Euclid's criterion of comparability between homogeneous magnitudes – was also a source of predictable perplexity. The so-called axiom of Archimedes, on which this criterion rested, theorized that any magnitude, even a very small one, when added to itself a sufficient number of times, can generate an arbitrarily large magnitude. In this sense, the infinitesimal was not 'Archimedean', since its use in calculus implied that multiplying it by an arbitrarily large finite number did not alter its nature as an entity near zero: an infinitesimal multiplied by any finite number still produces an infinitesimal.

Such anomalies could only confirm the 'fictional' character that Leibniz attributed to infinitely small quantities. Nevertheless, it was clear that the infinitesimal involved a more direct method of considering and resolving the problem which the techniques of Eudoxus and Archimedes would have described more rigorously. But even Archimedes, Leibniz thought, probably conceived his findings in terms of infinitesimal magnitudes, and then translated them into the language of the potential infinite. The practical advantages of differential calcu-

---

11. Ibid., p. 14.

lus were such that one could definitively translate it into Archimedean techniques. The utility of the innovation and the awareness of this equivalence sufficed to render tolerable the conceptual ambiguities implicit in this revolutionary mathematical exercise. Yet the whole truth of the matter cannot be expressed by saying that the entire sense of the infinitesimal was limited to its practical convenience. For Leibniz, infinitesimal fictions were *well-founded fictions*; and the concrete references that they necessarily made to natural reality, in which the infinite constantly came into play, lent an even greater degree of likelihood to their functionality as a calculating device. The conceptual ambiguities involved in their use were offset by their applied success as a calculus whose mechanisms so closely resembled those of nature. 'The infinites and infinitesimals are so well-founded,' Leibniz could write, 'that everything in geometry, and in nature as well, proceeds as if they were perfect realities.'[12]

The differential was an analytical correlative, manipulated in the spirit of Cartesian geometry, to the infinite subtlety of reality. Forces, velocities and stress could all be described as essentially instantaneous entities, and thus involved infinitesimal quantities. The infinitesimal was the description of a movement or a variation in its germinal state, and therefore a kind of quintessence or archetype of form, whose manifestation it anticipated as an invisible impulse. (*'Vis autem derivativa,'* Leibniz wrote, *'est ipse status praesens, dum tendit ad sequentem seu sequentem praeinvolvit.'*[13])

By virtue of its internal productivity and infinite fecundity, Leibniz's monadic substance was a real correlative to the purely ideal potential of the geometrical fiction. The predicates of a substance were infinite like the terms of a converging sequence, whereas the magnitude of a substance was finite like the sum of the sequence.[14]

Form and geometrical figure thus had the nature of units viewed from the perspective of an infinite series of component terms. The process of

---

12. Leibniz, Letter to Varignon, 2 February 1702; cited in R. Guénon, *Les principes du calcul*, p. 44.
13. 'The force of the derivative is the present state itself, as far as it tends to the following state or induces it.' See Brunschvicg, 'Spinoza', p. 52.
14. Ibid., pp. 53–4.

integration could then be regarded as an ideal transition to an effective completion of form, realized not as limit, but as a synthetic recapitulation of an infinite multiplicity, and ultimately as an actual infinite. Every form in nature was the visible result of an analogous mechanism.

It is hardly surprising that the certainty of seeing the infinite present and operating everywhere eventually prompted the idea of designating it by an algebraic sign. And this was probably the most revolutionary innovation.

Descartes had already offered an image of the infinite as a clear and distinct entity, listing its paradoxes as natural and inevitable features in something incompatible with the ordinary perception of limit. His words make it clear that he refuses to contemplate the infinite through authentic dialectical mediation, or to symbolize it as the limit point of any mental process which is based on hypotheses and *contradictions* or on the repeated *negation* of viewpoints aimed at establishing the central truth of an object. The cognitive value of contradiction, as revealed in Plato's *Dialogues* and recently revived by Karl Popper, is replaced by a decided tendency to regard even the unlimited as a form that is accessible to the mind and analysable with the instruments of reason.[15] In several of his replies to Mersenne's objections, Descartes' tone makes clear that the paradoxical nature of infinity no longer justifies proclaiming its non-existence. The contradictions of dialectic reveal *what* the infinite is, rather than revealing their inadequacy in understanding the idea. Even when experienced negatively as the absence of limit, the infinite becomes something explicit and *positive*.[16] The infinite finally loses its primal identification with negation and non-being, the qualities that made it the irrational (*alogos*) *par excellence*, the unnameable without remedy.

Still, in the writings of Descartes and other mathematicians before Leibniz – Kepler, Galileo, Fermat, Pascal, Cavalieri and his student

---

15. Cf. Karl Popper, *Scienza e filosofia* (Turin, 1969), p. 42, where Popper declares that the *falsifications* of theories indicate 'the points where we have, as it were, touched reality'.

16. Descartes, *Premières réponses aux objections*, in *Œuvres et lettres*, p. 353. [Cf. *The Philosophical Writings of Descartes*, trans. J. Cottingham, R. Stoothoff and D. Murdoch, vol. 2 (Cambridge, 1983), p. 81. Translator's Note.]

Mengoli – the mathematical use of the infinite did not yet appear as an authentic *mechanism*, as a calculation reproducing in different symbols the automatic operations of algebra.

The infinite had indeed been present and intuitable in the mathematics that preceded differential analysis. And even if they were still unknown during this period, the proofs of Archimedes' *Method* had already offered a recognizable image of it, but without compromising its character as *logos alogos* (irrational ratio). But the Greeks chose not to grasp or accept the advantages of algebraic generalization. Instead, it was Leibniz who discovered and exalted them. 'What I like best in the new calculus,' he wrote to Huygens in 1691, 'is that it offers us the same advantages over the ancients in Archimedean geometry that Viète and Descartes offered us in the geometry of Eudoxus and Apollonius: it relieves us of working with the imagination.'

Léon Brunschvicg traced the conceptual dependence of Leibniz's invention on the spirit of Descartes' philosophy and mathematics. Beginning with the Cartesian school, 'the notion of the infinite is as simple, as clear, and as distinct as that of the finite, whose algebra allows us to develop analytical properties according to the order of reason. After Pascal, there is a geometry of the infinite which cannot be reduced to Cartesian geometry, but its principles cannot be explicitly stated in the language of understanding. Instead, they are the object of *sui generis* intuitions, and their conclusions scandalize common sense and the rational reason of logicians. Leibniz's aim was to apply the form of Cartesian analysis to the geometry of Pascal, to the geometry of indivisibles. After Archimedes, or at least after Cavalieri, the procedures of integration no longer remained to be discovered; nor did the procedures of differentiation after Fermat. But there still remained the necessity of encompassing these procedures within the unity of an intelligible system ... Precisely where Pascal found his eyes closed as if by a sort of fatality, Leibniz discerns the possibility of a new generalization that itself will imply the need to make explicit what intuition implied, and to translate infinitely small elements into analytical symbols. At this point, Leibniz enters the school of Descartes.'[17]

---

17. Brunschvicg, 'Spinoza', pp. 49–50.

Did mathematics continue to use Leibniz's symbols? In fact, the notion of the infinitesimal remained in mathematical language and, even after Weierstrass, was still being used as a didactic fiction. Yet, before the work of Skolem (1934) on non-standard models of Arithmetic, and before the proof of the 'compactness theorem' (as Tarski baptized Malchev's findings of 1936), no one ever maintained that infinitesimals were based on solid conceptual clarity.[18] In 1942, Alonzo Church wrote that he found it preferable to introduce infinitesimals into his teaching 'in an openly imprecise and heuristic manner as small values of increment, or little bits of the implied variable quantity, rather than to dress up the idea with the deceptive appearance of logical precision'.[19]

But it is important to note the fact that the infinitesimal continued to be *used* in certain sectors of the mathematical discipline and in applied practice, although (as Church makes clear) its explicitly fictional nature tended to discredit it before Skolem and Robinson consolidated it. Leibniz's concessions about the purely ideal nature of infinitely small quantities, which might at any time have been suppressed out of respect for the more rigorous descriptions of Archimedes' method, ended up as the principal alibi for their survival in Analysis up until Weierstrass's work. For Robinson, non-standard Analysis helped rewrite a history of calculus aimed at re-evaluating the infinitesimal. In his view, through their singular combination of verbal and practical flexibility with logical incoherence, Leibniz's apologetic remarks obscured the real intentions of several mathematicians who are traditionally considered pioneers in the nineteenth-century refoundation of Analysis. Cauchy, for example, often spoke of infinitesimals and constantly used them in crucial definitions. But it is not sufficiently clear to what extent he considered them a mere linguistic shorthand for more rigorous formulas, rather than an indispensable part of the

18. A brief survey of attempts to justify the infinitesimal from the standpoint of non-standard Analysis is found in the last chapter of Abraham Robinson, *Non-Standard Analysis*, 2nd edn (Amsterdam, 1974), pp. 260–82: 'Concerning the History of the Calculus'. An explicit reference to the use of the compactness theorem in reformulating the principle of continuity is found, for example, in J. P. Cleave, 'Cauchy, Convergence, and Continuity', *British Journal of the Philosophy of Science* 22 (1971): 29.
19. Alonzo Church, 'Differentials', *American Mathematical Monthly* 49 (1942): 391.

proofs and formulations of theorems. In his *Course on Analysis*, Cauchy tried to derive infinitesimals from the notion of 'variable', since in this way the potential nature of the infinite would survive intact. He wrote that 'when the successive numerical values of a variable decrease indefinitely to be smaller than any given number, such a variable becomes what is called *infinitesimal*, or an infinitely small quantity' (*Cours d'analyse* [Paris, 1821], p. 4). And Cauchy also used formulations explicitly invoking the infinite to define the continuity of a function, introducing a formula that would later prove of great efficacy in teaching: a function $f(x)$ is continuous if infinitesimal variations of x create infinitesimal variations in f. We find the same notion of increment in his definition of a derivative. In addition, Cauchy's theorems, formulated in terms of the infinitely small, proved exact when translated into the language of non-standard Analysis, where infinitesimals conceived as an increment of variables were replaced by infinitesimals as perfectly defined entities in Robinson's expansion of the field of real numbers. Cauchy's famed 'erroneous' theorem, concerning the continuity of the limit function of a convergent series of continuous functions, is one of the cases that show how an 'abbreviated' description in terms of non-Archimedean quantities was not merely a linguistic convention, but an essential condition in formulating the problem. A translation in purely finite terms of Weierstrass's 'ε – δ approach' would have explicitly involved the notion of 'uniform convergence' as explained later by Weierstrass, and would have seemed an unlikely project on Cauchy's part.[20]

'Thus,' Robinson concluded, 'Cauchy stands in the history of calculus not as a man who broke with tradition . . . but rather as a link between the past and the future. The elements of his theory can be traced back to Newton, Leibniz, and beyond, but he provided a synthesis of the doctrine of limits on the one hand and of the doctrine of infinitely small and large quantities on the other by assigning a central role to the notion of a variable which tends to a limit, in particular to the limit zero.'[21]

---

20. On Cauchy's erroneous theorem and Weierstrass, see J. P. Cleave, 'Cauchy, Convergence, and Continuity', and Robinson, *Non-Standard Analysis*, pp. 273–7.
21. Robinson, *Non-Standard Analysis*, p. 276.

The gradual eclipse of the infinitesimal culminated in Weierstrass's work, but after him various attempts to salvage it were made by Du Bois-Reymond, Stolz, Schmieden and Laugwitz, and Skolem. In the end, Robinson used Skolem's studies of non-standard models of Arithmetic and stated in no uncertain terms that Leibniz's fictions *existed*. In other words, by virtue of a theorem of the logic of predicates (the so-called compactness theorem), there existed an extension of the ordinary field of real numbers **R** which he called *****R**, including infinitesimals and infinites. This was a proof that Leibniz had never developed, even if he was clearly on the right track in the practical application of his infinitesimal procedures.

But what *existence* is actually meant? Significantly, the proof of the compactness theorem implies a sort of extrapolation to infinity; and its use of *ultrafilters* and *Zorn's lemma* (or the *axiom of the choice*), which were famously contested by intuitionist mathematicians, implies a decisive assertion of the actuality of the infinite. As Wittgenstein suggests, the *sense* of a result cannot be read solely by the *proposition* that describes it, but must be read above all by the *demonstration* and by the new conceptual edifice that it inevitably erects. 'In mathematics,' he writes, 'we are convinced of *grammatical* propositions; so the expression, the result, of our being convinced, consists in the act that we *accept a rule*.' Hence, 'nothing is more likely than that the verbal expression of the result of a mathematical proof is calculated to deceive us with a myth.'[22] 'What is exact in mathematics,' Goethe asked himself, 'if not exactitude?'

The logical foundations of the infinitesimal, moreover, rest entirely on the calculation of predicates and on quantification. And it would be hard to devise something that illuminates the ontological implications of a science better than the ordinary logic of predicates as governed by universal and existential quantifiers. But in recognizing objects associated with the values of the variables bound by an existential operation, we would be wise to recall the cautionary statements made

---

22. Wittgenstein, *Bemerkungen über die Grundlagen der Mathematik* 26; Italian translation, *Osservazioni sui fondamenti della matematica* (Turin, 1971), p. 102. [English version, *Remarks on the Foundations of Mathematics*, trans. G. E. M. Anscombe (Cambridge, MA, 1978), p. 162. Translator's Note.]

by W. V. Quine in his *From a Logical Point of View* and *Word and Object*. The recognition of certain entities by a theory, Quine explains, is merely a consequence of their suitability, when replacing quantified variables, for satisfying certain propositions of that theory, that is, for causing the propositions to be *true*. (One might add that they are true in the sense described, for example, by Alfred Tarski in his article 'The Concept of Truth in Formalized Languages', in *Logic, Semantics, Metamathematics* (Oxford, 1956), pp. 152–268.) Yet everything that would seem to have acquired the right to *be* is immediately shifted by Quine to the plane of supposition and discursive fiction. In *Word and Object* (Cambridge, 1960), which obviously could not take note of Robinson's more recent work, the infinitesimal is treated as a conscious and deliberate myth, a fecund but fictitious shorthand for Weierstrass's most rigorous formulas. This singularly indulgent evaluation was clearly vindicated six years later in Robinson's *Non-Standard Analysis*. But concerning the *existence* of such infinitesimals, we could also make use of another earlier observation of Quine: in this case in *From a Logical Point of View*, where, speaking about *existence* on the topic of the quantified logic of predicates, he says it is preferable not to take for granted the *states* of things that objectively exist, but rather to try to clarify the ontological commitments of a discourse. The primary aim is to *understand* what is said to exist in a coherent language, without having to accept as a consequence (of its coherence) that its existence is real.

In *Meaning and Necessity* (Chicago, 1947, p. 42), Rudolf Carnap cited Quine's proposition: 'The ontology to which one's use of language commits him comprises simply the objects he treats as falling within the range of values of his variables.' But in Carnap's judgement, the term 'ontology' was already so compromised that it should be avoided. While declaring his substantial agreement with Quine, he refused to make even the slightest concession to the existence *per se* of the objects to which a linguistic framework refers. It is especially valuable to read Carnap if we wish to gauge the impropriety of investigating the 'real' existence of mathematical objects, and if we wish to grasp clearly the climate that was created by the scientific convictions which accompanied their development. He swept away the last metaphysical residue from mathematics, and regarded the traditional controversies between realism and nominalism as pseudo-problems lacking any cognitive significance. Above all, he carried to its limit the pragmatic vision of mathematical truths that Bernard Bolzano had initiated a century before. In the propositions of mathematical logic, what counts is not the demonstration that certain things exist, but the plausible proof that they are worth examining. Bolzano wrote that the introduction of absolute propositions and truths into logic should depend on their *utility*. Carnap wrote that the choice of a linguistic framework and of the types of variables should be considered a pragmatic decision, as if dealing with an engineer's instrument. The meaning of existence was thus linked to the specific use of the verb 'to be' within the language examined. As Quine too explained: 'To be is to be the value of a variable.'

Besides, any absolute declaration about existence or non-existence always risks appearing pointless. We can gauge the weakness of some lines of argument – caused by confusion about the priority of abstract names over real forms – when we tackle both the problem of universals (classes, functions, numbers, relations) and the case of all possible fictions. The most intransigent nominalist – one who (usually) accepts a quantified logic in which the values of variables range over a domain of unmistakably abstract objects – gains little credence when he affirms their existence. Nevertheless, it happens that nominalism objectifies universals with an ease unthinkable for a realist, and that in extreme cases it even considers universals as single concrete objects.[23] On the other hand, the realist formula *universalia ante rem* reinforces the thesis that the phrase 'there are' referring to abstract objects differs from the phrase referring to sensory data, and might suggest an analogous arrangement for fictions. So one may suspect that existential operators give fictions the same factual charge that countless examples would seem to confer on them even beyond any formal specification. Besides infinitesimals, one may think of *Don Quixote* or *Hamlet*, of Marx's reply to Kant's refutation of the ontological proof of St Anselm,

---

23. See José Ferrater-Mora, 'Fictions, Universals, and Abstract Entities', *Philosophy and Phenomenological Research* 37 (1977): 353–67.

of the disintegrating power of the duplications of Tlön narrated by Borges in his *Fictions*, or of certain formulations of the *Tao te Ching*. All of this apparently reinforces the traditional view of the priority of *Nāman* over *Rūpa*, of name over form.

10

The Principle of
Indiscernibles. Classes

In Musil's *Man Without Qualities*, Clarisse writes a letter to Ulrich
that offers a perfect image of our human folly when we try to realize
the actual infinite *in concreto* through an analytical and exhaustive
description of a set of objects. A newspaper account of a railroad
collision in Pennsylvania gives Clarisse a very simple idea. The locomo-
tive engineers did not purposely cause the disaster. Instead, they were
deceived by their ignorance of an entire network of events which were
in varying degrees responsible for the crash. 'That monstrous network
of tracks, switches, and signals that covers the whole globe drains our
conscience of all its power. Because if we had the strength to check
ourselves just once more, to go over everything we had to do once
more, we would do what was necessary every time and avoid the
disaster. The disaster is that we halt before the next-to-last step.'[1]

In fact, Clarisse cannot resist the perverse temptation of feeling
herself able to take this last step, which would be a daunting task even
for someone endowed with a keen memory or a disciplined mind, if
not for someone fortunate enough to be a genius. On the spot, Clarisse
mentally reviews a series of associated ideas that reveal who, of all
people, should be the depository of such genius: 'For all these reasons,'
she explains, 'I consider it my duty to meet Moosbrugger.'

When Clarisse in fact visits the murderer Moosbrugger in a criminal
insane asylum, she is the one who, unwillingly, is forced by apparently
random circumstances to stop at the 'next-to-last step'. After an

---

1. Robert Musil, *The Man Without Qualities*, III.7; Italian translation, p. 690. [English
translation by Sophie Wilkins (New York, 1995), p. 774. Translator's Note.]

interminable prelude of encounters with increasingly crazed and dangerous patients, the doctor who escorts her is urgently called away, and the true purpose of her visit is never accomplished.

Why does Clarisse fail to meet Moosbrugger? The fact is that Moosbrugger is one of those extreme cases that not even the legal system deigns to consider. He is capable of envisioning the boundless interdependence of earthly things, and therefore realizes that nothing can be detached from the whole because everything is linked to everything else. When psychiatrists show him a picture of a squirrel, Moosbrugger admits that he sees a fox, a hare, a cat, or perhaps something else. He knows that in different countries a squirrel is variously called a 'tree kitten' or a 'tree fox'. Clearly, 'if a tree kitten is no cat and no fox and no hare . . . you don't have to be so particular about what you call it; you just know it's somehow sewn together out of all those things and goes scampering over the trees.'[2] This is in fact the source of Moosbrugger's mild-mannered but bloodthirsty criminality. A person who fails to distinguish things and calls something indifferently by one name or its opposite – for even distant events are undeniably linked by some invisible affinity – cannot recognize what is good and what is evil, since both can be merged and neutralized in a sort of primordial, chaotic simplicity.

When Clarisse foolishly seeks to grasp the complexity of an event like the train crash, her ambition finds its ideal counterpart in Moosbrugger's introspection. But in Moosbrugger the awareness of an infinite network of interrelations is achieved at the price of a complete amorphousness and apathy, which can only be compared with the reciprocal indifference of the particles in the primordial chaos described by Anaxagoras and Anaximander. Thus, his homicidal instinct arises from an intolerance of any complete form that aspires to permanence, and it seeks to reduce everything to primal nothingness.

The evil of the infinite dominating the experience of Clarisse and Moosbrugger is shared by some of Dostoevsky's characters. Dostoevsky's 'man from the underground' is an emblematic character and the founder of a long literary tradition, in whom the error of

---

2. Ibid., II.59; Italian translation, p. 231. [English translation, p. 259.]

the infinite leads not to murder, but to grim brooding and to the disintegration of natural and orderly thinking. The same thoughts and heightened awareness that correlate distant and elusive facts may initially cause inertia and apathy, a passive but conscious state of paralysis in a muddle of doubts, torments, agitation and confused sensations.

The man from the underground justifies himself as follows: 'All spontaneous people and men of action are active because they are dull-witted and limited. How can this be explained? This is how: as a result of their limitations they take immediate and secondary causes for primary ones, and in this way they are more quickly and easily convinced than others that they have discovered an indisputable basis for their activity, and so they rest assured beforehand that there are absolutely no remaining doubts. But how am I, for instance, to reassure myself? Where are the primary causes that I am to take my stance upon, where are my bases? Where am I to take them from? I practise thinking, and as a result any primary cause I have immediately drags another in tow, one that is even more primary, and so on *ad infinitum*. And this is precisely the essence of any kind of consciousness or thought process.'[3]

The dazzling literary images of Musil and Dostoevsky inevitably found an echo in scientific and philosophical awareness. While the language and implications change, the substance of the discourse is similar if not identical.

Hegel wrote in his *Logic* that any determinate and finite being exists in relation to the entire world that surrounds it. Hence, as Paul Feyerabend explains in *Against Method*, the complete description of any object is a self-contradictory notion.

A passage in Alfred North Whitehead's *Science and Philosophy* describes the correlation between finite and infinite in terms of the complexity of an event, implying that one can never completely exhaust the list of connections that necessarily bind an entity to the totality

---

3. Dostoevsky, *Notes from the Underground*, Chapter 5; Italian translation, *Ricordi dal sottosuolo* (Florence, 1964), pp. 26–7. [English translation by Jane Kentish (Oxford, 1991), p. 19. Translator's Note.]

that contains it. 'The contrast of finitude and infinity,' he writes, 'arises from the fundamental metaphysical truth that every entity involves an indefinite array of perspectives, each perspective expressing a finite characteristic of that entity. But any one finite perspective does not enable an entity to shake off its essential connection with totality. The infinite background always remains as the unanalysed reason why that finite perspective of that entity has the special character that it has. Any analysis of the limited perspective always includes some additional factors of the background. The entity is then experienced in a wider finite perspective, still presupposing the inevitable background which is the universe in its relation to that entity.'[4]

Since we cannot know all the experiences which are linked to an object and make it an imperfectly calculable limit point, full of indiscernible nuances and aspects, any claim to affect reality through action and interference proves vain, or at best imprecise and risky. Musil dwells on the definition of this concept. When Diotima reproaches the sterility of Ulrich's conduct, he reminds her that a person's thoughts must form an autonomous entity, one independent and distinct from life, and that 'all one can do is refuse intellectual participation in reality.'[5] Of course, this does not *a priori* rule out the success of an action, but it remains something that at heart resembles the success of an artist, something that 'laughs at calculations', and is therefore indissolubly linked to the irrational, as Musil makes clear. Strikingly, *The Man Without Qualities* views the emergence of pervasive irrationality, dominated by the idea of a false infinite, from the perspective of a secret accord with the imperative to act. Diotima and Arnheim symbolize the two poles of this accord.[6]

---

4. A. N. Whitehead, *Science and Philosophy* (New York, 1948), pp. 85–6; Italian translation, *Scienza e filosofia* (Milan, 1966), p. 88.
5. Musil, *Man Without Qualities*, II.66; Italian translation, p. 264. [English version, trans. Wilkins, p. 296. Translator's Note.]
6. This point of view finds an exact parallel in *Tao te Ching* 29, cited from the Italian translation (Milan, 1973), p. 83: 'Those who seek to take command of the empire by action I have seen fall into an impasse. One cannot fashion the sacred vessel of the empire. Whoever fashions it, ruins it; whoever lays hold of it, loses it.' The reference to all action must then reside in the invisible centre of oscillation produced by the inevitable alternation of more and less, or limit and unlimited: 'Hence, sometimes things lead,

Does this bear any relation to mathematics? In point of fact, we often find in mathematics an archetype of spiritual issues; and the problems it discusses, under a different name, can reveal a surprising correspondence to the facts of life. 'If we translate "scientific outlook" into "view of life", "hypothesis" into "attempt", and "truth" into "action", then there would be no notable scientist or mathematician whose life's work, in courage and revolutionary impact, did not far outmatch the greatest deeds in history.'[7] Hence, 'in science it happens every few years that something till then held to be in error suddenly revolutionizes the field, or that some dim and disdained idea becomes the ruler of a new realm of thought. Such events are not merely upheavals but lead us upward like a Jacob's ladder.' And Musil also writes: 'In mathematics, we have a new method of thought itself, the very wellspring of the times and the primal source of an incredible transformation.'[8]

To measure the purport of these words, we need only ponder the historical significance of Leibniz's differential calculus, say, or the justification for Bertrand Russell's bold declaration that the methods of traditional philosophy had been rendered inadequate by recent innovations in mathematics.

The new perspective discovered by Thomas Kuhn in his *Structure of Scientific Revolutions* can be shifted to mathematical research, where it can help us understand the actual dimensions of certain changes. It is clear that innovative propositions do not by themselves create the new systems of rules that characterize the next scientific paradigm. 'The existence of a paradigm,' Kuhn wrote, 'need not even imply that any full set of rules exists.'[9] Moreover, 'scientists work from models acquired through education and through subsequent exposure to the literature often without quite knowing or needing to know what

---

and sometimes follow; sometimes breathe easily, and sometimes gasp violently; sometimes are vigorous, and sometimes distressed; sometimes begin, and sometimes decline.'
7. Musil, *Man Without Qualities*, I.11; Italian translation, p. 36. [English translation, p. 37.]
8. Ibid., p. 34. [English translation, p. 35.]
9. Thomas Kuhn, *The Structure of Scientific Revolutions*, 2nd edn (Chicago, 1970), p. 44; Italian translation, p. 66.

characteristics have given these models the status of community paradigms. And because they do so, they need no full set of rules.'[10] And the agreement of mathematical insights with accepted norms of ethics and behaviour – in short, with the entire 'grammatical' structure of life – reflects the fact that 'scientists . . . never learn concepts, laws, and theories in the abstract and by themselves. Instead, these intellectual tools are from the start encountered in a historically and pedagogically prior unit that displays them with and through their applications.'[11]

Wittgenstein taught that mathematical paradigms inevitably derive from a set of laws governing existence and ordinary praxis, which is substantially indefinable and lacks any absolute theoretical foundation. We could apply this doctrine to the reversal of perspective that we see in a direct comparison of Aristotle and Leibniz. Could anyone deny that, for the Greeks and European civilization after the Renaissance, a *mathematical* conception of the infinite signals the recognition of the radical distance and incompatibility between mankind's psychological, moral and even perceptual conditions?

It is well known that Leibniz continually associated his mathematical inventions with the language that he used to describe his vision of the world. In various ways, he linked souls with mathematical points; and to illustrate the structure of reality, he used geometrical imagery governed by the principle of continuity. At least within the limits of analogy, reality was assimilated to the mathematical continuum; and contingent existence, with its infinite complexity, was compared to the unfathomable irrationality of the points of a continuous geometrical space. In accordance with his intuitive notion of geometrical continuity, he said that nature 'does not make leaps'. In the apparent void between two distinct substances, it allows the existence of an infinity of other intermediate substances which differ from each other by infinitesimal degrees.

Thus, all created substances were arranged in a continuous series in which necessary and contingent realities occur in the same ratio as rational and irrational numbers. The reduction of rational numbers

---

10. Ibid., p. 46; Italian translation, p. 68.
11. Ibid.; Italian translation, p. 69.

to a common unit of measure was comparable to the possibility of explaining fundamental truths by means of identical verities. The infinity of irrational numbers, in turn, offered a parallel to the analysis of contingency, which is fatally unlimited and without solution. Only God can know and achieve the complete explanation of the causes behind an event.

If it is true, as Whitehead maintained, that the continuum offers the principal modality for representing indefinite potentiality, we would not be far from the truth if we imagined Leibniz's world as dominated by the unlimited. This is so even in the relatively precise sense that both it and its mathematical apparatus lack the idea of *limit* as an element dominating a potential infinity while situated *outside* it. From this perspective, all contingent reality consists of points obtained by continual variations through a problematic infinity of intermediate conditions and the infinitesimal steps between them. Only God is external to the world, transcending the unlimited series of monads. In this sense, we may at least perceive an analogy between the revolutionary character of infinitesimal mathematics and the world as conceived using the idea of infinity which Aristotle had rejected.

Yet in this picture of the world, the one feature which is not derived from the unlimited is diversity. Any two points of a continuous series, if distinct, must differ by some predicate that reveals their dissimilarity. But two points that have the same predicates and are therefore indiscernible must be the same point. By this principle of indiscernibles, Leibniz denied the existence of duplication, or the mirror-image opposition of two identical but distinct objects, since such a condition would paralyse the world, making every movement and choice absurd and contradictory.

But in order to formulate the principle of indiscernibles, we must first understand what it refers to. Whenever we speak of identity, and assert that two objects are identical if and only if they have all the same properties, we risk involving non-existent entities, or falling unawares into the circularity of logical or semantic paradoxes.

Bertrand Russell stripped bare the logical fabric of Leibniz's arguments, revealing their ambiguities and contradictions and demonstrating that only the methods of mathematical logic were suited to the correct reformulation of the principle of indiscernibles. The discussion of this question by Whitehead and Russell in *Principia Mathematica* showed that while Leibniz only dimly understood the infinite implicit in his discussion of indiscernibles, his mathematics could address problems of philosophy with utterly reliable criteria and proofs. Russell's intellectual attitudes – revealed especially in the more popular explanations of the *Principia*, and in his digressions on the history of philosophy – show how thinkers of a certain period tended to consider mathematics an instrument of absolute knowledge, capable of illuminating the most impenetrable concepts, including that of the infinite, by means of deductive chains derived from some elementary and seemingly indisputable logical concepts. But clearly no 'scientific' explanation of the infinite could be proposed which indicated exactly what it is in unambiguous terms. Instead, the result was the reinforcement of a certain agnostic attitude, in which there clearly loomed a hint, if not an admission, that the question should only be treated in the terms established by the new literature.

In any case, such logico-mathematical investigation refined our understanding of the idea of the infinite, and at the same time continually pointed to aspects of nuances of it similar to much older conceptions.

The logical theory of *types*, first formulated by Russell, was an antidote to the naive conception of a 'collection' and the paradoxes that it implied. The existence of such paradoxes was imputed to a vicious circle implicit in particular infinite classes of objects. Some of these classes had to be considered logically 'impossible' simply because their definition posited that they were already defined entities.

In a certain sense, the logical construction of *Principia Mathematica* reformulated the question: can any set form a totality when discussions of it are based on a blind faith in its existence? Quine justly observed that the chief aim of the *Principia* was to reduce the universe of classes to logical coherence, and this meant using a hierarchy of logical types that prevented the over-extension of instantaneous groupings of objects. For example, Whitehead and Russell discovered that it is not possible to consider the simultaneous totality of a group of individuals together with their attributes and relations, and with the attributes and relations of these attributes and relations, and so forth indefinitely. This 'indefinitely' necessarily entails an increasing complexity, such that it orders the universe into successive classes of logical meanings across a well-defined hierarchy.

Among the 'non-existent' classes, there is one that directly engages the principle of indiscernibles and the impossibility of correctly indicating all of an individual's infinite interrelations with things in the universe around him. A notion such as 'all the properties of object a' – meaning 'all the propositional functions that are verified by argument a' – is completely illegitimate, and necessitates a hierarchical ordering of successive classes of attributes. Among such classes of attributes, a central position is occupied by the so-called 'predicates' that indicate, as it were, the collection of properties 'closest' to the individual considered in the hierarchical order of types. Of course, predicates do not exhaust all the properties; and if we tried to found the principle of indiscernibles directly on *all* the properties of two objects x and y, concluding that only their complete equality would allow us to derive the identity $x = y$, we would face an insurmountable obstacle.

Naturally, the idea of identity requires a totality without compromises, since we tend to think that x and y are identical only if *all* the interrelations of x and y with the universe containing them are equal to each other without exceptions. But Russell demonstrated that identity in this total sense is logically non-existent and undefinable. By contrast, in a stricter sense it is possible to reformulate Leibniz's idea with the aid of the so-called *axiom of reducibility*, which in a technically specifiable sense restores an enunciative function to a predicating function. This reformulation enables us to describe the idea of identity as follows: x and y are identical only if x and y have the same predicates.

Yet the reduction of the monstrously unlimited universe of classes was not the only consequence of the discovery of paradoxes or mistrust of the unlimited. The very notion of class – an idea capable of instantly uniting an infinity of objects without subjecting it to the temporal becoming involved in the process of counting – was not adopted as a necessary and irreducible logical element. Whitehead and Russell write: 'The symbols for classes, like those for description, are in our system incomplete symbols: it is their *use* that is defined, and one assumes that they themselves have no meaning ... Therefore, the classes as we introduce them are pure symbols or linguistic conveniences and not authentic objects like their components, in the case that these are individuals.'[12]

A suggestive case that confirms this idea is the possibility, advanced in the *Principia*, of visualizing a class as 'the substratum of all the equivalent assertions concerning various formally equivalent functions'.[13] In other words, a class resembles something which we need not conceive as existing, but which is in some way determined by all the propositional functions that are verified by the same arguments, and that for this reason designate it; it is one of the hypostases that our imagination cannot help introducing and accrediting.

But the drastic reduction of consistent linguistic formulas effected by the *Principia* did not meet with universal and definitive approval. It is indisputable that at the time a new paradigm of research necessitated a clear reformulation of the ontological implications of language; and Russell's *theory of descriptions* offered the first significant antidote to the elimination of pseudo-problems of existence. Still, the suppression of classes did not reach the heart of the problem. Russell had to offset this reductive elimination by appealing to other universal entities, namely, to propositional functions or attributes. In so doing, he invited the criticism of Quine, who in substance considered this sort of reduction superfluous. Quine's perplexity was also directed at the stratification of types, which he considered in several respects as unnatural and as introducing clearly undesirable corollaries.

In the later axiomatic systems of Zermelo, von Neumann and Gödel-Bernays, a mistrust of the idea of *class* appeared in a more nuanced form. Zermelo's system definitively asserts the *existence* of certain sets, even if the constructivist approach sketched in his theory presents some arbitrarily limiting aspects which, in order to avoid contradictions, would exclude perfectly legitimate and unobjectionable sets. Von Neumann sought to eliminate the suspicion that the idea of class was in itself contradictory. Instead, he suggested, it is the use made of some

---

12. Whitehead and Russell, *Principia Mathematica* (Cambridge, 1973), p. 72.
13. Ibid., p. 74.

classes as *elements* of other classes that creates logical inconsistency. If some of Zermelo's axioms moved within the context of constructible sets, by means of a formulation of well-defined predicates, *within* the groupings of approved legitimacy, von Neumann reintroduced some of the more 'dangerous' definable classes, not as subsets of something else, but by directly (as it were) using predicates that could be extended to the entire universe. Such was the *universal class*, which was already involved in some of the principal antinomies, or paradoxes of set theory, and which corresponded to the predicate '$x = x$', whose final and conclusive nature would be defined, in von Neumann's view, by its exclusion as an *element* in any collection of objects.[14]

Von Neumann's formulation of the question was adopted by Kurt Gödel and Paul Bernays. Their axiomatic system distinguishes between *set* and *class* by saying that a set can be an element of a class, while a class is not an element of anything. The *existence* of classes is explicitly formulated by Bernays in a meta-mathematical theorem that can be proved by using the first axioms of the elementary extensionality and constructibility of sets and classes, and that predicts the definability of groupings of objects by predicates of the most general type, thus establishing a precise correspondence between classes and predicates.[15] In their *Algebra* (New York, 1967), Saunders MacLane and Garrett Birkhoff used Bernays' system of axioms as the theoretical foundation for their algebraic theory of *categories*, which implies classes of objects characterized by a high degree of abstraction and *extension*.

This possibility of indefinitely expanding and widening the active domain of mathematics to levels of arbitrarily large generalities, while at the same time restricting algebraic discourse with well-defined limits immune to the antinomies, is clearly represented by the notion of a *Universe* (or *inaccessible Universe*). All the possible logical difficulties related to new categories created on the basis of given initial sets can be exorcized in each case by defining a 'container' set (or *Universe*), from which it is impossible to 'escape' by effecting the constructions

14. See Abraham A. Fraenkel, *Foundations of Set Theory* (Amsterdam, 1958), p. 97.
15. See Paul Bernays, 'A System of Axiomatic Set Theory', *Journal of Symbolic Logic* 2 (1937): 65–77.

envisioned by Bernays' axioms. We might regard this as a new approximation of the absolute, in which each step reproduces the features of the actual infinite which Cantor perceived in other mathematical entities, but which still unfolds in the context of a purely potential infinite.

In the first half of the twentieth century, re-evaluations of the idea of class or grouping suggest that the eventual problems of consistency had to be considered in the realm of semantics, and therefore required a careful re-examination of the use of *language*. The rethinking of these issues stimulated by the celebrated findings of Löwenheim and Skolem proved exemplary. In 1915, Löwenheim showed that, in the calculus of first-order predicates, every finite set of properly defined formulas based on an infinite model must always have a countable model. In 1920, Skolem extended this conclusion to *infinite* countable sets of formulas, and later freed his proof from the burden of the axiom of choice, and revealed how these theorems could be derived from Gödel's findings about the completeness of the calculus of first-order predicates. In short, this meant that the interpretability of a system *L* of axioms – such as those of von Neumann or Zermelo and Fraenkel, for example – could always be limited to a *countable* domain of objects that verified them. But a difficulty arose as soon as one considered Cantor's theorem on the non-countability of the '*power* set' of natural numbers – meaning the set of all the subsets of these numbers. The theorem was perfectly demonstrable in the system *L*, but seemed at the same time to imply sets whose elements could not in any way be *counted* one at a time, and placed in a one-to-one correspondence with the series 1, 2, 3 . . . of natural numbers.

How could such sets fit into a purely countable model of the system? This paradox, known as *Skolem's paradox*, could provoke renewed mistrust of the infinite, especially if it crossed the boundaries of pure countability. An interpretation proposed by Skolem, and later adopted by Bernays, led to the conclusion that 'in reality everything is countable, and differences of power occasionally exist, but only in appearance, as it were, and relative to a conceptual picture.'[16] Skolem's solution

---

16. Cited in Ettore Casari, *Questioni di filosofia della matematica* (Milan, 1976), p. 124.

meant that Cantor's theorem had to be considered an assertion of the *non-existence* of a particular set. This set was the one representing the one-to-one correspondence between the set of natural numbers and its power set, and formed by ordered pairs of elements from the first and second sets. But this non-existence was not *absolute*. It was limited to the range of possible models of set theory which could be derived from a consistent group of axioms. *Outside* this theory – that is, in a meta-language – the countability of the continuum could be envisaged.[17] A set, it was concluded, was capable of existing or not existing, depending on the linguistic perspective from which it was viewed.

As Ettore Casari observes, the essential point was the intrinsic fallibility and imprecision of the axiomatic instrument as a descriptive system, the irremediable imperfection of language formalized according to the usual schemes of axiomatics.[18] Borel observed that Cantor's theorem had revealed the existence of an indefinite component of the continuum; and his observation may prove most meaningful within this linguistic gap, like the broader question of the existence of determinate infinite classes. Skolem's so-called relativism, which is an inevitable consequence of his paradox, deprives the notion of 'class' of what might be called its absolute character. The signs or words that represent it belong to a partly illusory text that was to furnish an exemplary paradigm for that undefined moment in history when the identity of the signifier and the signified had presumably faded into illusion.

---

17. See Rudolf Carnap, *The Logical Syntax of Language* (New York, 1937).
18. Ettore Casari, *Questioni di filosofia della Matematica*, p. 124.

II

The Actual Infinite. Indefinite
and Transfinite

How did the more recent belief develop that the infinite can exist as an actual totality? A list of precursors would prove endless. Going back through the centuries, we have seen that we would find Jean Mair, Gregory of Rimini and Duns Scotus. More recently, a description of the infinite in Spinoza's twelfth letter, and Hegel's commentary in *The Science of Logic*, would offer us the most perfect expression of a possible alternative to the unlimited conceived as pure potentiality.

The systems of Spinoza and Hegel maintain that the infinite must necessarily be static if it is to cease being a mere synonym of temporal becoming. Even while believing in the actual infinite, Leibniz would perhaps not have insisted on this point to the same extent. It is true that the potential infinite in series finds a natural conclusion in the finite and thus in something solid and static. But Leibniz's emphasis is wholly placed on movement, and on the instant of its germination. As Gaston Milhaud remarked, the differential is simply the 'infinitesimal moment of every becoming'.[1] And the finite is the final act of a dynamic generation governed by time. Thus, when a series of alternating signs (whose partial sums are alternately bigger and smaller than the final sum) reach a final equality, Leibniz regards this equality as a dynamic entity captured in the infinitesimal as it passes from motion to rest. To define it, Fermat coined the new term *adequality* (*adégalité*) instead of *equality* (*égalité*). In fact, equality can be considered as a very small inequality, and each term may be made as close to the other as we wish.[2]

---

1. See Léon Brunschvicg, *Les étapes de la philosophie mathématique* (Paris, 1929), p. 209.
2. Ibid., p. 208.

It is useful to return to Descartes once more, if we wish to trace the idea of an infinite that can be translated into a mentally accessible notion with a name or symbolic designation like everything else in the world. Indeed, Descartes had already declared optimistically that our conception of the infinite is not unlike that of a finite figure. 'Just as it suffices for the possession of an idea of the whole triangle to understand that it is a figure contained within three lines, so it suffices for the possession of a true and complete idea of the infinite in its entirety if we understand that it is a thing which is bounded by no limits.'[3] In fact, Descartes never went so far as to admit the possibility of 'comprehending' the infinite and its entirety: rather, the impossibility of doing so reflected the 'formal cause' of the infinite and was implied in the truth of the idea that one could form of it. If we can conceive 'something', we cannot conceive the infinite, because the infinite transcends any 'thing' conceived. Nevertheless, we are aware of this, and this awareness reveals the existence and the truth of the idea. In this way, Descartes had at least proposed an infinite warranted by a legitimate content of objectivity, even if he did not consider it referable to any concrete thing.

These apparently harmless observations on the inaccessibility of the infinite – at least as an idea expressed in an ordered set of words and phrases – were in fact a warning that even the infinite, like anything else, could become the object of a mechanical invention, the unlimitedly reproducible sign created by an art that Leibniz was to define as feasible without using the imagination. Descartes was well aware that, whenever we adapt a definition to a practical application, we must still preserve all its original sense. Yet it is clear that the risk of understanding an object is always a sort of dissipation, an unlimited expansion of multiple exemplifications governed by laws that are no longer necessarily conceived *ab initio*. It was clearly also for this reason that the Pythagoreans banished from their school the man who first dared to divulge the mystery of incommensurability, the *alogos* or irrational *par excellence*.

---

3. Descartes, *Réponses de l'auteur aux cinquièmes objections faites par M. Gassendi* in *Œuvres et lettres* (Paris, 1953), p. 491. [English version, 'Author's Replies to the

In the course of the eighteenth century, Leibniz's students faced the task of expanding and consolidating the original design of Leibnizian doctrine, which they achieved not without triumphal exclamations and posturing inspired by rigid dogmatism. Despite the caution of Leibniz, who had kept the label of 'fictions' for differentials, Fontenelle wrote in the preface to his 1727 *Éléments de la géométrie de l'infini* that infinites and infinitesimals must now be considered as a fact established for all time, as an inevitable conquest of geometrical speculations. Descartes' concession that the unlimited did not exist as a really verifiable totality remained unchanged, even though the reality of the infinite was translated by *fiat* into a mathematical fiction. Fontenelle wrote: 'Geometry is entirely intellectual, independent of any actual description and of the existence of the figures whose properties it discovers. Everything that geometry conceives as necessary is real relative to the reality that it supposes in its object. The infinite that it demonstrates is thus just as real as the finite, and its idea of the infinite, just like other ideas, is an idea of supposition which is merely convenient, and which must vanish as soon as it has been used.'[4]

In any case, the interminable discussions that took place in the eighteenth century about the metaphysical foundations of the differential and the principle of continuity did not alter the conviction that a plausible idea of the infinite must in some way be affirmed and consolidated in mathematical language. The infinitesimal had clearly not lent itself to an unimpeachable logical foundation, but at the practical level it suggested the idea of undoubted solidity. It is not surprising, then, that we find the existence of the infinite, at least as an intellectual reality, accepted as a firm conviction in the man who was often considered the most direct precursor of Cantor's work, Bernard Bolzano.

In the first half of the nineteenth century, Bolzano reached conclusions not unlike those already proposed by Descartes, but at the same time he hazarded decidedly more daring theses. In his *Wissen-*

Fifth Set of Objections', in *The Philosophical Writings of Descartes*, vol. 2 (Cambridge, 1984), p. 254. Translator's Note.]
4. Brunschvicg, *Les étapes*, pp. 243–4.

schaftslehre (*Theory of Science*) 48–9, he began by writing that ideas in general are not something 'existing', but simply 'something' that does not cease to subsist even when no one thinks about them. There are also ideas which deserve the name of 'objective ideas', as being endowed with a clearly unequivocal quality, but to which no concretely existing object corresponds. Such are the ideas of 'nothing' or '$\sqrt{-1}$' or 'o' (zero). Bolzano concluded that the determination of a thing does not necessarily require its effective existence. And the infinite was no exception: while it is true that an infinite set cannot be defined by an exhaustive description of its elements, there are still methods that allow us to define it unambiguously.

But Bolzano was not content with recognizing a correctly determinable objectivity in the idea of infinite. Not unlike Leibniz, he saw the infinite as operating everywhere in reality. In his *Paradoxien des Unendlichen* (*Paradoxes of the Infinite*) 25, he linked the infinity of God to the infinity of created beings and with the infinity of their experiences, however brief, and even ventured the thesis – a bold and dubious one, to say the least – that the infinity of God can only be based on multiplicity. From this point, it began to be assumed that mathematicians would become the legitimate trustees of any clear interpretation of the concept of infinite multiplicity.

A decisive innovation by Bolzano, who in this respect even surpassed Cantor and directly foreshadowed twentieth-century trends in research, was his outline of a propositional logic capable of reformulating the problems of *existence*, which nearly anticipate the results of Russell's *descriptions* or Quine's efforts to minimize the ontological commitments of language. In his *Theory of Science* 142, Bolzano makes clear that he distinguishes between different types of existential propositions. To say that '*A* exists' is not the same as saying that 'there is an *A*', because existence appears as a predicate in the first proposition, but not in the second. The formula 'there is an *A*' simply refers to the fact that the *idea* '*A*' is referential, and this permits a use of the verb 'to be' that is exempt from ontological compromises. Thus, the proposition 'there is a supreme moral law' does not mean that something really exists that is a supreme moral law (*Theory of Science* 137). And the ambiguity lies not only in the propositional form, but in the intrinsically problematic nature of ideas like 'supreme moral law' that refer to something that does not exist, even if it is still 'something'.

It is not absurd to conceive the referential content of an idea as relating to something non-existent. It suffices to recall that the idea of a 'proposition' can refer to things like the Pythagorean theorem or the theorem of the parallelogram of forces. Such theorems do not exist as actual realities, even considered as *propositions in themselves*: for we need not suppose that someone thinks or writes them at a given moment (*Theory of Science* 49). The *truth* of such propositions does not lie in their reference to concrete objects that behave 'in reality' as they describe them. The Thomist formula 'true and being are convertible terms' (*Verum et ens convertuntur*: *Summa theologica* 1.16.3) leaves room for ambiguities when we deal with objective ideas whose reference lacks any spatial or temporal existence and yet which enter into some relation with the verb 'to be'. As Bolzano writes in *Theory of Science* 4, when we say that a truth says how things *really* stand, we must recognize that we are speaking in a figurative sense. For it often happens that neither being ('*Sein*'), nor actual presence ('*Dasein*'), nor existence ('*Existenz*') nor actuality ('*Wirklichkeit*') belong to the proposition that describes the truth or the things to which it refers. A linguistic formula like 'there is an *A*', in which existence is not an explicit predicate, can then subsist beyond any ontological commitment and can assume a sense close to that inferred from existential quantifiers in the ordinary logic of predicates.

For all of Bolzano's insistence on situating the infinite among *real* things (in *Paradoxes* 25), he prepared the ground for the advent of a *harmless* use of the verb 'to be' with reference to universals and fictions. He neutralized, as it were, the evocative power of the sign; and in the spirit of the most radical nominalism, proposed that it could be reduced to manipulable *fiches*. In this way, the infinite could have appeared in language without excessive ontological commitments, and been subject more to the needs of pragmatism than to the suggestions of metaphysics. One of the most peculiar characteristics of traditional symbolisms would thus have been lost. Half a century before Bolzano, Novalis had written that every word is an evocation, and that the spirit evoked each time *appears*; and this happens by virtue of the intimate bond which governs the general sympathy of the universe, and which is expressed in purest form in the mathematical sign.[5] This implicitly postulated a realm of correspondences at the level of the invisible, ready to manifest itself, sometimes still enveloped in invisibility, in the texture of the subtlest analogies, sensations, emotions and consequences that the sign evoked in more or less recognizable form. Hence, we might define Bolzano's innovation as an exorcism of the spectres called up by the symbol, a kind of lubrication of logical thought effected by suppressing certain *invisible* components of a proposition or idea. The pairing of Novalis and Bolzano is obviously one of many possible ones; but it offers a notably recent outline of the typical opposition, found throughout the history of the relation between being and sign.

In the second part of his *Philosophy of Symbolic Forms*, Ernst Cassirer sees the earliest moment of identifying being and word in passages of the *Rigveda*. If we went back to the most perfect image of the word soaring to the level of invisible being, it would mean recalling the Hindu *Vāc*, who in the *Bṛhad-āraṇyaka-upaniṣad* couples Hunger with Death to engender everything that exists. In Heraclitus' aphorisms and in Plato's *Cratylus*, we find two of the most remarkable examples of the correspondence between name and thing. The *logos* of the Greeks is analogous to the Hindu *Vāc*: in its profound complexity, it embraces the correspondence of signifier and signified in its double aspect of revelation and veil, of concrete allusion and hidden or illusory truth. Novalis' utterances revitalized this perspective; but neither Novalis nor other modern thinkers like Vico or Johann Georg Hamann (who reintroduced the Heraclitean vision) could prevent the total unity of being and language, implicit in the mythical image of the world, from the collapse that began after the Renaissance.

In *Les mots et les choses*, Michel Foucault traces the essential lines in the evolution of the symbol from the sixteenth century onwards. He suggests that the emergence of a purely *binary* organization of symbols, especially in the *Logique de Port-Royal*, constitutes a fundamental overturning of the sixteenth-century vision of a world criss-crossed by innumerable correspondences, marked by the 'similitudes' known as

---

5. Novalis, *Fragments* 1153; Italian translation (Milan, 1976), p. 295.

analogia, *sympathia*, *aemulatio* and *convenientia*. In the seventeenth and eighteenth centuries, we witness the gradual dimming of *similarity* as the third link between analogous signs, and as a principle of indefinite expansion of correspondences. Each symbol now remains alone facing its own meaning, in the restricted space of its exclusively referential or semantic content. This short circuit of the symbol – abstracted in its bare function as the binary relation between signifier and signified – gave rise to successive theories of epistemology conceived as the study of *representation*: empiricism, ideology, and the investigations of Locke, Hume, Berkeley, Destutt de Tracy and Condillac. The theory of knowledge as the elaboration of a total system of signs emerged as a general analysis of every form of thought using ideas drawn from the senses and from abstraction, and fostered a two-fold development: a science of nature in a semi-dogmatic key, and a philosophy of representation destined to become increasingly sceptical and nominalistic. Bolzano comes after Locke, Hume, Hobbes and Condillac, but he adds to them what might be called a singular mixture and synthesis between Platonic realism (a term which, while still in use, is inadequate) and the 'devitalization' of the idea. This devitalization ultimately occurs in a universe of signs which increasingly resembles a formalized system or language that analyses, breaks down and re-assembles everything in a combinatory game separate from the realm of being.

The 'real' existence of ideas and propositions was drastically reduced by Bolzano to their eventual appearance in some point of space-time, to their being thought of or uttered in a determined moment – obviously without considering the possibility of a pre-existing spirit 'materializing' at that moment. And this is one of the essential points. From that time on, the sphere of magic, which Novalis had been among the last to claim even for mathematical signs, indeed especially for them, remained hidden in silence. Novalis himself had written that man is closely bound to what is invisible, far more than to what is visible. Whether he likes it or not, man is bound to an invisible world which, in his naivety or in an unconscious attempt at exorcism, he tries to rule out as 'non-existent' and unable to influence events.

Bolzano's reductive gesture could also be explained by citing his speculations on the concept of truth. To avoid misunderstandings, he explains in *Theory of Science* 30, the verb 'to be' as we use it in speaking about a *truth in itself* has a wholly conventional sense; it must be separated from the idea of actuality, and of an operative presence that conditions the facts of life. The truths of religions, morality, mathematics and metaphysics are only propositions or ideas to be regarded in the first instance as things in themselves, distinct from the *fact* they describe, which generally 'does not exist'. 'True propositions' resemble the fictions of geometry, and their truthful nature lies in the internal law of their composition, rather than in their similarity to facts (*Theory of Science* 24–5). In this 'neutral' terrain, cleared of pseudo-problems, an infinite that was finally 'devitalized' and abstracted as a concept stripped of any set of possible correspondences might well have found a reassuring environment for surviving as an intelligible sign of rational thought. Its evocative power was already nearly ruled out as a game of phantoms, and as mere metaphysical stupidity. In this culmination of the catastrophic evolution of the symbol, perhaps for the first time language was completely protected from the incursions of *being*.

In other words, the symbolic density of the sign, together with its network of correspondences at various levels of being, had now been definitively transposed to the sundered and separate space of the *poetic* sphere, while the language of science retained only a rigid system of similarity relations. This system lacked the validation of a unique and primordial Text wherein the infinite resemblance of things was described; and it bore no relation to Foucault's 'primary, invisible writing' of the sixteenth century, compared to which every imaginable language seemed a mere commentary.

Bolzano advanced the proposal that a class of objects was definable without reference to its being finite or infinite. He anticipated Cantor in maintaining that one should verify the idea of set by means of the Aristotelian principle of the 'excluded third'. A thing is determined or determinable if only one of two contradictory attributes (b and non-b) applies to it.[6] This rule is equally applicable to both finite and infinite. Even if it is infinite, a set is determinable because each of its elements

---

6. Bolzano, *Theory of Science* 87.

does or does not possess a specific property. Like Cantor later, Bolzano based his faith in the existence of the actual infinite on the conviction that, if a mental process could assemble separate and distinct objects into a higher unity, it was immune to logical contradictions.

Later, this belief was disproved by the facts, so that the attempt was abandoned to provide a general and immediate definition of the concept of set. In any case, Bolzano's ideas preceded by several decades the period of the nineteenth century that was marked by the discovery of a mathematical language aimed at expressing the infinite as an actual and static totality no longer governed by temporal becoming. This was the language of the theory of sets, which aimed at justifying, besides the potential or syncategorematic infinite of infinite series, the actual infinite of all the subsets of these series as a closed collection of objects existing by themselves.[7]

Bolzano anticipated this development by demonstrating that sets are formed by synthetic atemporal operations. When we speak of the set of the inhabitants of Peking, we individuate a well-defined set without being required to enumerate separately all its components one by one. Analogously, the terms of an infinite sequence can all be specified by the law governing the formation of the sequence, which renders superfluous enumerating its terms: it is the law that specifies the sequence, and not 'all' its terms counted one by one.

Weierstrass's 'ε – δ approach' reflects the same needs for stasis. If we state that a function $f(x)$ is continuous at the point $x = a$, we mean that $f(x)$ converges to $f(a)$ when x becomes arbitrarily close to a. But this convergence, which suggests the idea of becoming and potentiality, can also be described in the following terms. For *every* positive quantity ε, there *exists* a corresponding positive quantity δ such that, for *all* real numbers x that are found in a neighbourhood of a with a radius δ, the value of $f(x)$ is found in a neighbourhood of $f(a)$ with a radius ε. In other words, $a - \delta < x < a + \delta$ implies that $f(a) - \varepsilon < f(x) < f(a) + \varepsilon$. The key terms of the definition ('every, exist, all') suggest infinite and static totalities which logically precede the actual choice of points contained

---

7. See Hermann Weyl, *Philosophy of Mathematics and Natural Science* (Princeton, 1949), p. 46.

in them, and are independent of their eventual management (by some mathematical law) as a potentially infinite sequence, and hence as a sequence subject to the inexhaustibility of temporal becoming.[8]

Weierstrass was also the first to found a theory of irrational numbers on the notion of 'set', extending operations and relations of magnitude to infinite aggregates of rational numbers. The entire arithmetic of real numbers was thus susceptible of description in terms of sets. The potential infinity of the figures formed by irrational numbers was enclosed in an actual entity governed by laws independent of any idea of temporal succession.

Later, Cantor would explicitly touch on the essential point, namely, that this viewpoint principally affected the idea of time. The belief that Kant's aprioristic concept of time was unimpeachable suffered a categorical refutation in the most explicit manner. (The aprioristic concept of space had collapsed with the discovery of non-Euclidean geometries.) Thus, concerning the relation between continuity and time, Cantor established the clear priority of the idea of the continuum. 'I must affirm above all,' he wrote, 'that in my opinion introducing the notion of time or the idea of time must not serve to explain the more primitive and general notion of the continuum. In my opinion, time is an idea which, in order to be clearly explained, presupposes the notion of continuity independently of the notion of time. Furthermore, with this notion of continuity time can be conceived neither objectively as a substance nor subjectively as a necessary *a priori* idea. This idea of time is merely an auxiliary and relative idea, useful in establishing the relation between different movements that take place in nature and that we perceive. Thus, nothing ever appears in nature that resembles objective or absolute time, and consequently we cannot adopt time as a measure of movement; on the contrary, movement could be considered as a measure of time if we weren't hindered by the fact that we gain nothing by considering time as a subjective idea that is necessary *a priori*.'[9]

---

8. Ibid.
9. Cantor, 'Fondements d'une théorie générale des ensembles', *Acta mathematica* 2 (1883): 403.

The crisis of Kantian apriorism sketched in these words played a decisive role in mathematics during the late nineteenth century, especially when we consider that it was the development of an arithmetic and geometry outside of time which made possible a definition of the infinite in static terms, one that was ultimately susceptible of being transformed into a mathematical foundation of the actual infinite.

Ernst Cassirer observed that nineteenth-century mathematics witnessed a progressive dimming of the cognitive value of the forms of sensible intuition. Dedekind, Russell, Frege and Hilbert followed Cantor in seeking to reduce the foundations of number to logical constants and to primary and autonomous relations of thought.

Dedekind thought that even the continuity of space could derive from these same constants. The process of the arithmetization of mathematics, to which his works contributed decisively, hinged upon the idea of number that Frege had regarded as the property of a *pure concept*, rather than of a sensible thing. Summarizing this viewpoint, Cassirer writes: 'The series of numbers is not to be built on the intuition of space and time; it is just the reverse; the concept of number, an "immediate emanation of the pure laws of thought", can alone enable us to gain precise concepts of space and time.'[10]

If at this point we wished to find the exact antithesis of this viewpoint, it would be useful to turn back to one of the last witnesses of the archaic conception of number, such as that of Simplicius: 'Time is the number of a certain movement or, better and more generally, the interval proper to the nature of the Universe.'[11] There is no better introduction to this statement, and to all its implications, than the arguments that Giorgio de Santillana employs in *Hamlet's Mill* to demonstrate the existence of an inseparable union between *number* and *time* in archaic thought.

While strident and disorienting, such parallels have the merit of showing us a perspective in which the criterion of truth that seems to inspire Cantor's statements becomes less self-evident. Later, Poincaré,

---

10. Ernst Cassirer, *Philosophie der symbolischen Formen*; Italian translation, *Filosofia delle forme simboliche* (Florence, 1961), p. 218. [*Philosophy of Symbolic Forms*, vol. 1, trans. Ralph Manheim (New Haven, 1953), p. 227. Translator's Note.]

11. See Pierre Duhem, *Le système du monde*, 10 vols. (Paris, 1954–73), vol. 1, p. 80.

Brouwer and Weyl were to find serious reasons for refuting Cantor, and to this end would use witnesses even more ancient than Simplicius. Duhem was in a position to observe that certain pages of Damascius, the teacher of Simplicius, would not be out of place in the writings of Bergson and (we might hypothetically add) in those of Brouwer.[12] Cantor's certitudes were thus attacked by those who returned directly to the theses that he had hoped to destroy using the evidence of mathematical results. The intuitionism of Weyl and Brouwer gave contemporary Kantian apriorism a new lease on life.

Yet before that happened, mathematicians found arguments that decisively refuted the Aristotelian and Thomist theses concerning the non-existence of the infinite. The true turning point came with Cantor's theory of sets at the end of the nineteenth century, but some mathematical discoveries anticipated it by closely reproducing the mechanisms for generating transfinite numbers and uncountable sets. In particular, a theorem of du Bois-Reymond, of which an account is given in Émile Borel's treatise *La théorie des fonctions* (Paris, 1928), is worth noting.

Let us imagine that we choose a sequence of increasing functions $f_1(x)$, $f_2(x)$, ..., $f_m(x)$, ... For example,

$$f_1(x) = x,$$
$$f_2(x) = 2x,$$
$$f_3(x) = 3x,$$

. . . .

. . . .

$$f_m(x) = mx,$$

. . . .

In other words, $f_m(x)$ is the function that for every number x yields the product of x multiplied by m. This sequence possesses an evident property: as x varies in the range of positive integers, the values of $f_m(x)$ are less than the values of $f_n(x)$ whenever n is larger than m ($n > m$). We may succinctly describe this fact by writing

---

12. Ibid., p. 271.

$$f_m(x) < f_n(x) \quad \text{if} \quad n > m.$$

Thus, the functions $f_1(x)$, $f_2(x)$, ... form an *indefinite* sequence such that each element of the sequence is 'greater' than those that precede it in the sense defined by the preceding relation. In other words, we may write

$$f_1(x) < f_2(x) < \ldots < f_m(x) < \ldots$$

Is there an increasing function $f(x)$ which is 'greater' than *all* the functions $f_m(x)$? If our inquiry of this function remains tied to $\{f_m(x)\}$, the answer is obviously negative. But if our inquiry can be conducted outside the sequence, such a function exists in a certain sense. Let us think of $f(x) = x^2$, that is, of a function that for every number yields the square of that number. This function has the property of 'exceeding' any other $f_m(x)$ for any given m, at least after a certain x. In this sense, it 'grows' more quickly than *all* the functions $f_m(x)$, and in the same sense we can state that $f(x) = x^2$ is 'greater' than these. From this point, we could start a new sequence of functions:

$$g_1(x) = f(x) = x^2,$$
$$g_2(x) = x^3,$$
$$g_3(x) = x^4,$$
$$\cdot \quad \cdot \quad \cdot \quad \cdot \quad \cdot \quad \cdot \quad \cdot$$
$$\cdot \quad \cdot \quad \cdot \quad \cdot \quad \cdot \quad \cdot \quad \cdot$$
$$g_m(x) = x^{m+1},$$
$$\cdot \quad \cdot \quad \cdot \quad \cdot \quad \cdot \quad \cdot \quad \cdot$$

Here too, if we imagine x running through the series of positive integers, we clearly have

$$g_m(x) < g_n(x) \quad \text{if} \quad n > m.$$

There exists, then, a function $g(x)$ which is 'greater' than all the functions $g_m(x)$ without belonging to the sequence $\{g_m(x)\}$. In fact, it suffices to suppose that $g(x) = 2^x$, since the exponential function grows 'faster' than any power x^m. The same function 2^x could now generate another indefinite sequence of increasing functions that can be augmented by a further function $h(x)$; and the function $h(x)$ thus derived could itself form the beginning of a new, analogous procedure. We would thus obtain a sequence of functions

$$f(x), g(x), h(x), \ldots$$

for which we could infer the existence of a function greater than $f(x)$, $g(x)$, $h(x)$ and *all* the subsequent functions.

In fact, the existence of functions exceeding a potential infinity of increasing functions is guaranteed for each specific case by a theorem of du Bois-Reymond. General in nature and rich in implications, the statement of this theorem may be formulated as follows: 'Given any countable sequence of increasing functions

$$f_1(x), f_2(x), \ldots, f_m(x), \ldots$$

of the real variable x there exists an increasing and effectively constructible function $f(x)$ such that $f(x) > f_m(x)$.'

Evidently, the theorem transfers to the rank of mathematical 'fact' the rule that underpins any valid attempt to exhibit the actual infinite as a concrete event: an entity capable of limiting a potential infinity of objects of which it is not a part. It is also true that the same theorem suggests the possibility of exceeding every imaginable limit, for it opens a perspective in which an apparent metaphor for definitively achieving the infinite is only a provisional boundary suggesting a continuation. As the example described suggests, it is clear that no countable sequence succeeds in filling up the class of functions that can be constructed using du Bois-Reymond's theorem. In fact, by the same theorem *any* 'indefinity' (that is, any 'countable infinity') of increasing functions admits a new, greater function outside itself. This simply means that the word 'indefinite' is inadequate for a synthetic description of the class of functions that can be generated using the mechanism suggested by the theorem. Not only is there always something beyond what we have succeeded in counting; there is also something that we shall never succeed in reaching, even if we postulate the existence of 'limit' entities that exceed the inexhaustible potential of counting. Thus, the 'insatiability' of *apeiron* passes from the 'indefinite' to the 'transfinite'. The condition by which every number is followed by another (which defines the sense of the term 'indefinite') is transposed to the analogous condition, which predicts that any indefinite will be followed by a limiting term that produces a new unit, and therefore a further countable infinity (which defines the sense of the word 'transfinite').

All the same, du Bois-Reymond never explicitly spoke of 'transfinity'. Instead, it was Cantor who derived a systematic theory of the infinite from the mathematical facts that hinted at it. He did this not by using the findings of du Bois-Reymond, which he knew, but by extending beyond their immediate significance the findings and theorems concerning the convergence sets of a trigonometric series. He enlarged the research in this field launched by Fourier, Hankel and Riemann, while adopting Weierstrass's viewpoint and approach, which were already set-theoretical and static, and hence directed towards the manipulation of actually infinite aggregates.

In 1883, Cantor could declare that the time was ripe for the introduction of a new kind of infinity. This had already happened *de facto* in geometry and in the theory of functions. The study of analytical functions of a complex variable presupposed that one could examine points on a plane at an infinite distance from the origin, where the functions' properties remained analogous to those at finite points. Now, the inquiry concerning the structure of the intervals of convergence of trigonometric sequences led to a consideration of systems of points to which the operation of 'derivation' was applied an infinite number of times.[13] The result was sufficient to conclude that beside the *improper* infinite – which was the potential or syncategorematic infinite of Aristotle and St Thomas, but also of mathematicians like Euler, Cauchy and Gauss – one could also legitimately conceive a *proper* infinite, one that was perfectly determinate and hence actual.

Cantor found a similar character of actuality in the numbers by which counting could be extended from the indefinite to the transfinite. It would probably not be unfair to relate his discovery, by a convenient analogy, to the findings of du Bois-Reymond, in which we may see a source of Cantor's inspiration, even if he engaged different problems.

---

13. See the preface by Philip E. B. Jourdain to Cantor, *Contributions to the Founding of the Theory of Transfinite Numbers* (New York, 1955).

A similar parallel was noted and described by Borel in his *Théorie des fonctions*.

We must bear in mind that du Bois-Reymond's theorem can be indefinitely applied to any countable infinity of increasing functions. This indefinite application is triggered by a unique function $f(x)$, which is increasing and greater than x at least when x is sufficiently large. For example, we could take $f(x) = 2^x$ and define the sequence

$$f_1(x) = f(x) = 2^x,$$
$$f_2(x) = f(f_1(x)) = 2^{2^x},$$
$$f_m(x) = f(f_{m-1}(x)) = 2^{2^{\cdots^{2^x}}},$$

. . . .

A function 'dominating' such a sequence could then be indicated by the symbol $f_\omega(x)$, and the relation

$$f_\omega(x) > f_m(x) \quad \text{for every } m,$$

could be written briefly, by pure convention, as

$$\omega > m \quad \text{for every } m.$$

In turn, the function $f_\omega(x)$ could generate a new countable infinity of functions

$$f_\omega(x)$$
$$f_\omega(f_\omega(x)) = f_{2\omega}(x),$$

. . . .

$$f_\omega(f_{(m-1)\omega}(x)) = f_{m\omega}(x),$$

. . . .

to which we may again apply du Bois-Reymond's theorem. By analogy with the preceding case, we could call the new 'dominating' function $f_{\omega^2}(x)$; and we could describe the fact that $f_{\omega^2}(x) > f_{m\omega}(x)$ for every m, by conventionally writing

$$\omega^2 > m\omega \quad \text{for every } m.$$

The function $f_{\omega^2}(x)$ could then be used to generate another countable infinity of functions. It is clear that, just as counting presupposes a successive integer for every integer, so every countable infinity of functions A is followed by a further countable infinity B that can be constructed by a 'dominating' function. Successive applications of du Bois-Reymond's theorem thus suggest a sequence of symbols

$$\omega, \omega^2, \omega^3, \ldots, \omega^\omega, \ldots, \omega^{\omega^\omega}, \ldots, \alpha, \ldots$$

which we may consider as successive indices of the functions as they are progressively defined. This sequence allows us to imagine an extension of the process of counting which in some way reflects the structure of the class of functions generated by using du Bois-Reymond's theorem.

It is clear that Cantor's transfinite numeration can be introduced in an entirely autonomous way. But its analogy with similar findings, such as those of du Bois-Reymond, lends it the appearance of legitimacy and truth which we usually associate with what seems to be a 'fact' or a mathematical 'proof'. Still, we are only dealing with a mere analogy. Cantor's transfinite numbers were justified by *a priori* principles, rather than mathematical theorems or facts. As in the case of the real numbers defined by Dedekind, it was a set of postulates and ultimately a free act of creation that determined the existence of these new objects.

At this point, it is best to use Cantor's own words, and to cite verbatim a decisive passage from his 1883 article: 'Now we must show how we arrived at the definitions of these new numbers, and how in the actually infinite series of real numbers we obtain the natural divisions that I call *classes of numbers*. The series (I) of real positive integers $1, 2, 3, \ldots, \nu, \ldots$ is generated by the repeated positing and uniting of units which are presupposed and regarded as equal; the number ν is the expression both for a determinate finite number of successive positings and for the uniting of the posited units in a unique totality. The generation of *finite* real integers thus rests on the principle of the addition of a unit to a number which has *already been formed*; I call the *first principle* of generation this moment which, as we shall soon see, plays an essential role in the production of higher integers. The number of the numbers ν of the class (I) formed in this way is infinite; and among all these numbers, there is not one that is greater than all the others. It would therefore be contradictory to speak of a greatest number of the class (I); still, we can imagine a new number, which we shall call ω, that *will serve to express the fact that the entire set* (I) *is created by its law in its natural order of succession.* We can also represent the new number ω as the limit towards which the numbers ν tend, on condition that we understand that ω will be the *first* integer that *will follow all* the numbers ν, such that we must declare it greater than *all* the numbers ν. By adding the original units to the number ω, with the help of the *first principle* we obtain the more extended numbers

$$\omega + 1, \omega + 2, \ldots, \quad \omega + \nu, \ldots$$

Since in this way we still do not arrive at a greatest number, we imagine a new number, which we may call 2ω and which will be the *first* after all the numbers ν and $\omega + \nu$ obtained thus far. If we then apply to the number 2ω the first principle of generation, we arrive at the extension of the numbers obtained so far, as follows:

$$2\omega + 1, 2\omega + 2, \ldots, \quad 2\omega + \nu, \ldots$$

'The logical function that has yielded the two numbers ω and 2ω is obviously different from the first principle of generation; I call it the second principle of generation of real integers, and define this principle more clearly by saying: *If there is given any definite succession of real integers, of which there is no greatest, on the basis of this second principle of generation we may suppose a new number which is considered the limit of the first numbers, that is, which is defined as the next greater to them all.*

'By the combined application of these *two principles* of generation, we successively obtain the continuations of the numbers that we have obtained so far, as follows:

$$3\omega, 3\omega + 1, \ldots, \quad 3\omega + \nu, \ldots$$
$$\cdots \cdots \cdots \cdots \cdots$$
$$\mu\omega, \mu\omega + 1, \ldots, \quad \mu\omega + \nu, \ldots$$
$$\cdots \cdots \cdots \cdots \cdots$$

'Even so, we have not arrived at the end, since among these numbers there is no one that is greater than all the others.

'The second principle of generation thus allows us to introduce a number which follows all the others $\mu\omega + \nu$, and which we can designate as ω^2. To this number we may refer, in a determined order of succession, the numbers

$$\lambda\omega^2 + \mu\omega + \nu$$

and following the two principles of generation, we obviously come to numbers of the form:

$$\nu_0\omega^\mu + \nu_1\omega^{\mu-1} + \ldots \nu_{\mu-1}\omega + \nu_\omega;$$

but then the second principle of generation leads us to define a new number that will immediately be greater than all these numbers and that can be designated as ω^ω.

'As we see, the generation of new numbers is *endless*; following the two principles of generation, we constantly obtain new numbers and new series of numbers, with a *perfectly determinate sequence*.

'*One could therefore believe* initially that we shall lose ourselves in the indefinite in this generation of new, determinate, and infinite integers, and that we are not in a position to *temporarily stop* this endless procedure, in order to arrive thereby at a *limitation like the one we have in fact found, in a certain sense, in relation to the preceding class of numbers (I)* . . .

'*But if we now observe that all the numbers obtained so far and those that immediately follow them satisfy a certain condition*, we shall see that this condition, *if we posit it as obligatory for all the numbers that are to be formed immediately*, appears to be a *third* principle, which comes to be added to the first two and which I call the *stopping or limitation principle*. By virtue of this principle, as I shall make clear, the second class of numbers (II), defined with the addition of this principle, acquires not only a more elevated power than class (I), but *precisely the next greater power*, and consequently the *second power*.

'As we shall be immediately persuaded, the aforementioned condition is satisfied by each of the infinite numbers α defined so far, and is this: *that the system of numbers found in the series of numbers before the one being considered*, beginning with 1, is of the same power as the first class of numbers (I) . . .

'We therefore define the second class of numbers (II) as the set of all the numbers α that can be formed using the two principles of generation, which follow each other according to a determined order:

$$\omega, \omega + 1, \ldots, \nu_0\omega^\mu + \nu_1\omega^{\mu-1} + \ldots + \nu_{\mu-1}\omega + \nu_\mu, \ldots,$$
$$\omega^\omega, \ldots, \omega^{\omega^\omega}, \ldots, \alpha, \ldots$$

and which are subject to the condition that all the numbers that precede the number α, beginning with 1, form a system of the same power as the class of numbers (I).'[14]

Thus the first transfinite integer ω must be understood as a limit to which the finite variable integer ν tends, in the same way as an irrational number such as $\sqrt{2}$ can be seen as the limit of a variable. In a certain sense, Cantor wrote, transfinite numbers resemble new irrational entities; indeed, they imitate their profound nature, since both outline the features and indispensable characteristics of the actual infinite.

If we extrapolate Taylor's observation about the visualizing of irrational numbers as Platonic syntheses of the One and the Two, we may compare Cantor's third principle of limitation to the theorem that establishes the unambiguous definability of Dedekind's sections. Both refer to a boundary and a limiting correction. The first produces a dam that stems the overflowing power of generation of the second principle, restricting it (like du Bois-Reymond's theorem) to applications with countable infinities. The second delimits the infinity of oscillations of which it is the indispensable centre of reference and the ideal stopping point. This continues to be a sign of the One acting on the Two, or as Simone Weil expressed it, a 'sign of the dominion of the infinite over the indefinite'.[15] In effect, it was Cantor's outstanding achievement to impose an ordered structure on the immeasurable complexity of the transfinite. By virtue of the principle of *limitation*, he identified a hierarchy of successive 'cardinality' within which one could group the

---

14. Cantor, 'Fondements', *Acta mathematica* 2 (1883): 385–8.
15. Simone Weil, *Cahiers*, vol. 3 (Paris, 1974), p. 41. [*The Notebooks of Simone Weil*, trans. Arthur Wills, 2 vols. (London, 1956), p. 426. Translator's Note.]

numbers generated by the first two principles. 'With this method,' he wrote, 'by following these three principles *we can always arrive at new classes of numbers, and by these arrive at all the different powers, successively increasing, that are found in material or immaterial nature*; the new numbers thus obtained always have the same concrete precision and the same objective reality as the preceding ones . . .'[16]

The analogies between Dedekind and Cantor also reveal the fundamental style of thought and the profoundest beliefs concerning the nature of mathematical invention. In his preface to, and in the first section of, *Was sind und was sollen die Zahlen?* (*What Are Numbers and What Should They Be?*), Dedekind did not scruple to declare that arithmetic evolves perfectly independently of *a priori* intuitions of space and time, and that the concept of number is an immediate result of the laws of thought. In a manner that recalls Bolzano as well as Cantor, he imagined that we can use the ambiguous noun 'thing' to describe any object of our mental activity, and he began to refer to the operative convenience of designating such 'things' by simple symbols drawn on paper.

Although such signs are *not* identified with the things themselves, but are merely conventional representations useful for briefly recalling designated objects, we may question this alleged neutrality of the written symbol. As for the archaic resemblance of signs and things, and their power to evoke analogies and to trigger movements of sympathy and correspondences at all levels of reality, it would be difficult to regard them as uprooted and erased, even, we may suppose, in Dedekind's interpretation. For Dedekind, a 'thing' could be completely determined by everything that can be thought or said about it. And this bold affirmation must reflect his belief that this same 'thing' could be designated by a name or sign. The more determinate an object appears, the more easily it can be named. But this custom can scarcely be taken for granted when it happens that the object of this alleged determination is infinity.

Cantor made similar statements. He wrote that integers, as well as transfinite numbers, can be considered 'actual' to the extent that on

---

16. Cantor, 'Fondements', *Acta mathematica* 2 (1883): 390.

the basis of definitions they 'assume a perfectly determinate place in our knowledge, are clearly distinct from all the other constituents of our thought, stand in definite relation to them, and therefore modify the substance of our mind in a definite way'.[17] Yet in some way the term 'actual' also involves the reality of the external world, in which Cantor senses that one can find a correlative to our intellectual inventions. In 1895, he prefaced one of his last publications with this Latin epigraph: '*Neque enim leges intellectui aut rebus damus ad arbitrium nostrum, sed tanquam scribae fideles ab ipsius naturae voce latas et prolatas excipimus et describimus.*'[18] In other words, we do not arbitrarily make the laws of thought or the world, but receive them, like faithful scribes, as they are announced by the voice of nature. Hence, as Cantor wrote in another place, 'integers with their laws and relations constitute a totality in the same way as celestial bodies.'[19]

Even so, Cantor did not dare venture an opinion on the effective 'existence' of the new mathematical entities. He separated mathematics from metaphysics, maintaining that the former was free to move and develop in perfect autonomy, subject to the sole condition of non-contradiction and the intrinsic coherence of its statements. The problem of 'existence' was metaphysical and as such extraneous to the aims of mathematics.

Hence, Cantor arrived at a position bordering on formalism, and at least around 1882 went so far as to assert that real numbers should be regarded as mere signs that initially lack any property and are only supplied *a posteriori* with a system of rules that indicate their sense. Cantor saw that even with transfinite numbers he could not furnish even an approximate understanding of the Absolute. The Absolute can only be acknowledged ('*anerkannt*'), but never known ('*erkannt*'), even approximately.[20]

This idea of *impossibility*, which one encounters surprisingly often even in the sphere of a rigorously mathematical mental procedure, for

---

17. Cantor, *Contributions to the Founding of the Theory of Transfinite Numbers* (New York, 1955), p. 67.
18. See Abraham Fraenkel, *Abstract Set Theory*, 4th edn (Amsterdam, 1976), p. 80.
19. Ibid.
20. Cantor, *Contributions*, p. 62.

the most part indicates the insurmountable obstacle of the potentiality connatural in any object of thought. Just as Aristotle had already observed, there is always a *beyond*, something *more*, that has not been calculated or foreseen. This impossibility sometimes acquires the sense of a 'non-definability' or of an 'incalculability'. In an article of 1938, Alonzo Church[21] pointed out that Cantor's second class (II) of transfinite numbers – even when characterized by the sort of objective reality that Cantor had attributed to it – contains elements that elude any constructive definition and any denomination which is the final step in an effective computation. Borel's suggestion – that Cantor's diagonal proof of the uncountability of the continuum should reveal its intractably indefinite nature – could be transposed to the so-called second class of transfinite numbers. In accordance with the sense of the term 'calculability' as defined by Kleene, Turing and Church, it was proved that elements of the second class existed that could not be described in the formulas of λ-formalism, by the simple fact that the notion of constructibility designated by these formulas could be extended in only a limited way to recursively enumerable sets.

Cantor had already striven to prove that the power of numbers of the second class was greater than what is countable.

---

21. Alonzo Church, 'The Constructive Second Number Class', *Bulletin of the American Mathematical Society* 44 (1938): 224–32.

12

The Antinomies, or Paradoxes
of Set Theory

After Cantor, there were some who tried to recapitulate the proofs and justifications of the mathematical use of the actual infinite. The most insistent note sounded was a call to liberate the notion of number from the intuitive and empirical support of the ordinary process of counting, to found number on a purely logical basis, and to liberate every mathematical definition from any requirement of verification in the world of sensible phenomena. Ideally, the system of arithmetic should find its truest fulfilment in an autonomous elaboration of *a priori* concepts. Every justification was to be sought in the internal necessity of laws operating in a purely mathematical construction.

The principles governing Cantorian and post-Cantorian mathematics, which sanctioned the extension of number to the transfinite, were not unlike the reasons which Gauss, for one, had adduced to extend number to include the imaginary.[1] Behind these reasons lay a tacit premise. The acquisition of new concepts could be succinctly justified in the same terms that Musil's young Törless hears when his mathematics professor defends imaginary numbers: 'You must accept the fact that such mathematical concepts are neither more nor less than concepts inherent in the nature of purely mathematical thought.'

Cantor had pointed out that the current use of the infinite in geometry allowed the introduction of the 'proper' or actual infinite beside the syncategorematic infinite of Aquinas. After Cantor, Louis Couturat

---

1. See Ernst Cassirer, *Teoria della sostanza e della funzione* (Florence, 1973), p. 78. [English version, *Substance and Function*, trans. W. C. and M. C. Swabey (Chicago, 1923), p. 55. Translator's Note.]

emphasized the intimate analogy between the geometrical infinite and imaginary numbers. In essence, both could be referred to mathematical entities that were more readily accessible to the ordinary imagination. With the appropriate transformations, the infinite could be related to configurations in which the finite alone played a part. And the imaginary elements of a figure corresponded to the real parts of a figure of the same kind.[2]

In Poncelet's *Traité des propriétés des figures* (1822), the *principle of continuity*, which was an extension of Leibniz's analogous principle, underpinned this kind of deduction: the actual infinite was a sort of corollary to it. But if we sought to find a reason that furthered the extension of arithmetic to the transfinite, we would emphasize the transformation of the idea of number, which included some references to antiquity. Commenting on Cantor's work, Couturat noted that one of his merits lay in extolling the unitary and 'organic' character of the integers. Unity is not only the constituent element of number and of plurality, but also the conclusive bond that guarantees its totality.[3] Number is essentially 'cardinality': it is a class of classes in one-to-one correspondence with each other. In order to exist, this abstraction needs no special enumeration, for enumeration is merely an 'explicit knowledge of a number already thought of and implicitly determined'. Number is not the result of counting, but rather a condition for it. Consequently, to form an idea of number we may dispense with knowing multiplicity as temporal becoming; and Arithmetic cannot be a science of time, as Kant had wished.[4]

Couturat developed Cantor's theses by making number a formal entity that is completely defined *a priori* and implies no material or concrete content. In view of the unity guaranteed by number, he employed a metaphor from traditional metaphysics, and suggested that the multiplicity of phenomena plays the role of a prism traversed by a ray of white light: it refracts and diffuses it into things.[5] Couturat's treatise *De l'infini mathématique* (1896) represents the culmination of

---

2. See Louis Couturat, *De l'infini mathématique* (Paris, 1973), p. 266.
3. Ibid., p. 348.
4. Ibid., pp. 352–3.
5. Ibid., p. 360.

THE ANTINOMIES, OR PARADOXES OF SET THEORY

this static conception of number and the actual infinite, which he presented as a truth inferred from Cantor's theorems and from the law of continuity as formulated by Leibniz and others.

Continuing into the twentieth century, analogous endeavours led to what Fraenkel defined as one of the most dramatic events in the history of mathematics: the first proofs of the theorem of 'well-ordering' proposed by Zermelo in 1904 and 1908. Although the case was controversial, Zermelo succeeded in demonstrating that in *every* set M, and particularly in every infinite set, we may define what is called *well-ordering*, or a relation of order, which for simplicity's sake we might imagine as analogous to the 'greater' or 'lesser' of numbers, and which is characterized by the following property. Each of the non-empty subsets of the given set M contains one element that is smaller than all the others. In particular, for *every* pair of elements a and b in M we can say which is the 'greater' or the 'lesser' of the two. (This is not true for the weaker property of *partial ordering*, where the question 'Which of two elements is smaller?' may remain unanswered, as in the case of the relation of inclusion between sets.) In fact, as Lebesgue noted in a letter to Borel, Zermelo was able to show, more precisely, that a set M can be 'well-ordered' whenever we admit the actual existence of a function that associates a particular element m. with *every* subset M' of M'[6] This function of *choice*, conceived as existing, transformed the inexhaustible potentiality of the ordering operation into a static and total reality.

The intuitive method of ordering, at which Cantor's preliminary attempts stopped short, had to be a sort of counting of the elements to be ordered one *after* another in temporal succession, based on an extension of the ordinary procedure of counting to the transfinite. The hypothesis of the existence of Zermelo's function of choice – which became an axiom of set theory – broke the bonds of the temporal succession of choices by rendering simultaneous elements that time had made successive.

Zermelo's findings were of great importance to mathematics. They made it possible to extend the principle of induction – and of transfinite

---

6. See Émile Borel, *Leçons sur la théorie des fonctions* (Paris, 1928), p. 153.

induction in particular – by applying it to a far greater number of cases than had been foreseen. They also made an indispensable contribution to a satisfactory theoretical foundation of cardinal numbers. The classes of infinite sets that were equivalent, or in one-to-one correspondence with each other – which Cantor had prudently called 'powers' to avoid applying the idea of number too hastily to such abstractions – could more reasonably rise to become 'numbers' if their comparability was proved in a general way to be the same as that of finite ordinary numbers. This comparability was guaranteed by Zermelo's theorem of well-ordering. Indeed, the theorem of well-ordering, the axiom that established the existence of the function of choice, and the theorem of comparability between 'powers' were all mathematically equivalent: each of these theorems implied the others. They were three analogous formulations of a single fundamental choice: the replacement of the potential infinite – which was traditionally associated with time, and to which Cantor continued to grant a parallel existence, calling it the 'improper' infinite – by the infinite conceived as an actual, static and atemporal totality.[7]

Still, Cantor's mathematical inventions did not escape criticism and stubborn opposition. Such criticisms went beyond the level of friendly disagreement about questions of method and procedures. In some cases, they even turned into violent quarrels and gave rise to profound enmities, to the point of compromising Cantor's physical and mental health. Until his death, Kronecker led the opposition against any attempt to establish the actual infinite, and went so far as to attribute Satanic qualities to Cantor's innovations. In fact, he followed a path that was the opposite of the one that had led to the principal innovations concerning the infinite: the definition of the continuum with the relative proposals of Dedekind and Cantor, and the extension of the counting procedure to the transfinite. Dedekind and Cantor approached the problem of the continuum in a way that implied a sort of 'intrusion', within the domain of the pure arithmetic of integers, of propositions

---

7. On the debate about the infinite that led to Zermelo's theorem, see the interesting views expressed in the letters of Borel, Hadamard, Baire and Lebesgue collected in Émile Borel, *Leçons sur la théorie des fonctions*.

or theorems that could only be proved at the cost of introducing new postulates. To propose real numbers as effective signs for the *measurement* of determinate quantities in geometry entailed introducing an axiom which established that a series of integers associated with irrational numbers could be assigned to certain *points* of space. For Kronecker, this represented a hybridization of arithmetic with intuitions proper to geometry. It was an artificial device meant to effect the problematic reconciliation of the science of whole numbers, which was a purely mental product, with the science of space, which was still conceivable as something *external* because it was in some way linked to sensible intuition. In fact, Kronecker aimed at establishing an arithmetical science that could be extended to include algebra and analysis, but was also clearly separated both from geometry and from mechanics.[8] In this sense, he predicted an 'arithmetization' based on number conceived in the strictest possible sense. As a result, even negative, fractional and algebraic numbers had to be related to particular expressions which could be defined using integers and the functions of indeterminates (*indeterminatae*) which Gauss had introduced. The 'existence' of algebraic numbers – meaning the roots of an equation obtained by setting a polynomial equal to zero – was viewed simply as the recognition of intervals in which the polynomial changed sign. Thus, by avoiding fractional numbers using indeterminates and congruences, one could refer this existence to integers alone.[9] For Kronecker, even the assistance of logic was to be regarded as extraneous to the contents of pure arithmetic. In his view, an argument's obvious logicality offers no guarantee that its use is legitimate. To demonstrate a theorem, it is not sufficient to prove that negating its formulation implies a contradiction. In general, only an effective procedure, based on a finite number of arithmetic operations, leads to a certain result.[10] This suffices to exorcize the danger of the antinomies, or paradoxes of set theory, and to delineate a style of research which would become that of the intuitionists.

8. Leopold Kronecker, *Werke*, vol. 3/1 (Leipzig–Berlin, 1899), pp. 249–74.
9. Ibid., p. 272.
10. See Philip E. B. Jourdain, 'The Development of the Theory of Transfinite Numbers (Part III)', *Archiv der Mathematik und Physik* 22 (1913).

The discovery of logical paradoxes, beginning with that of Burali-Forti in 1897, intensified the controversy and at the same time served to show that those objects which had been claimed to 'exist' by virtue of simple acts of mental creation were often corrupted by intrinsic contradictions.[11] In the early years of the twentieth century, the opposition grew, and was joined by mathematicians like Poincaré, who wrote in 1908: 'There exists no actual infinite (given in its totality). Cantor's followers have forgotten this, and have fallen into contradictions.'[12]

After Cantor, anyone who denied the existence of the actual infinite proposed by the new language of sets found it necessary to criticize something else that Cantor had in part inspired: the logicist theses of Russell and Whitehead's *Principia Mathematica*. The *Principia* and some of Russell's later comments implied in effect that some of Cantor's findings had both a certain quality of incontrovertible rigour, and the potential to eradicate deep-rooted philosophical beliefs in the name of the new mathematical evidence. This new evidence, moreover, was a direct consequence of the process of arithmetization of mathematics achieved by Weierstrass and Cantor. In this visualization, integers were sufficient to describe the results of Analysis. But integers had been described in Peano's symbolic logic, which seems to lay bare a primary logical fabric capable of providing a foundation for the more elementary process of counting. Through arithmetic, logic thus made a powerful contribution to the recognition of the infinite as totality. From simple logical premises, conclusions appeared to follow which traditional philosophy had usually rejected. Among these were a reliable basis for the actual infinite and a new definition of the continuum, which would eliminate once and for all the philosophical claims of those who asserted that points in space exist solely in potentiality. The

---

11. As an example of an antinomy, I cite Russell's celebrated example of 1902: let *A* be the set of all the sets *S* such that *S* is not a member of itself. Then *A* is an element of *A* if and only if *A* is not an element of *A*. The existence of sets that are elements of themselves is exemplified by the set of all the sets which, *qua* set, must belong to itself. This existence derives from the 'naive' definition of set: it is guaranteed for every multiple of distinct objects by any condition whatsoever.

12. See Irving M. Copi, 'The Burali-Forti Paradox', *Philosophy of Science* 25 (1958): 281–6.

new definition of the continuum was stated as 'absolute', completely *a priori* and free of any spatial-temporal bond. As a distinctive characteristic, it implied a structure composed of simple constituent elements, real numbers or points – entities which the philosophy inspired by Aristotle and others had always regarded as inadequate to represent the parts of a whole characterized by the idea of continuity.

In emphasizing such questions, Russell displayed a seemingly unyielding conviction. 'It was often held ... that any subject-matter possessing it [continuity] was not validly analysable into elements. Cantor has shown that this view is mistaken, by a precise definition of the kind of continuity which must belong to space.'[13]

Still, sentences like 'Cantor has shown that this view is mistaken' lead one to wonder whether Russell is formulating something more incontrovertible than a simple opinion. As early as 1905–6, Poincaré observed that the language of Peano's symbolic logic, which inspired so many of Russell's theses, offered certainties that were in part illusory.

The formulas of this language, which seemed to make no concessions to naive common sense, employed combinations of signs and symbols intended to create the impression of manipulating primary truths and facts. At the time, the 'definition' of such truths seemed within reach, since the game of the symbols expressing it bore the hallmarks of irreducible rigour and simplicity.

In a sense anticipating Wittgenstein, Poincaré pointed out that such formulas were inadequate to represent authentically primary facts. He uncovered the suppositions and concepts implicit in the language that was used to express their definition. And he showed how difficult it was to encapsulate what characterizes mathematical thought in a definitive linguistic formula.[14]

Outside of mathematics, it has always been easier to resist the temptation of regarding a linguistic formula as a verbal image of a truth. Any assertion has a relative value, wrote Nicholas of Cusa, and is susceptible of being refuted. We need only recall the dialectic of Zeno

---

13. Russell, *Principles of Mathematics*, p. 271; Italian translation, *I principi della matematica* (Rome, 1971), p. 438.
14. See Henri Poincaré, 'Les mathématiques et la logique', *Revue de Métaphysique et de Morale* 13 (1905): 815–35.

or Gorgias, or Plato's dialectic, in which (as Simone Weil observed) two essential ingredients, *contradiction* and *analogy*, are used to move beyond a 'point of view' and beyond the confirmation of anything that rigidly claims to present itself as a single and separate judgement.[15]

We infer the same lesson when we read Cervantes' *Don Quixote*. Everything in the work tends to render vain and futile any perspective from which reality might be solely contemplated. In Part 1, Chapter 45, when the real nature of Mambrino's helmet is discussed – is it a helmet or a barber's basin? – the problem appears insoluble. The obvious solution to the question is mixed up with the most unlikely hypotheses, and the illusory game that ensues hints at the truth as an ineffable limit-point. The man who thinks he possesses the truth, and claims that the helmet is a basin, quits the debate mocked and defeated. In *Don Quixote*, the linguistic description of reality is quickly transformed into a game of epithets and designations, in which a thing constitutes an invisible point of reference while its essence remains irremediably hidden.

By contrast, we are usually led to believe that mathematics is a specially privileged instrument for inquiring into certain concepts, even offering absolute definitions and definitive solutions. Yet the legitimacy of this act of faith has not always been immune to suspicions. For example, we know, from the notes of his student Yorick Smythies, that Wittgenstein spoke about Cantor's infinite in terms that scarcely fostered a belief in the actual infinite as something to be taken for granted. To a large extent, Cantor let himself be guided by the fascination and suggestiveness of the numbers he was inventing, and further insisted on how marvellous it was that a mathematician's imagination could transcend any limit.[16] In such cases, Wittgenstein implied, the scientist's *arbitrium* ends up in the kind of conviction which the apparent incontrovertibility of mathematical facts exercises over anyone who is prepared to convert his passion for science into idolatry. Scientific findings readily endorse assertions of the type 'This *in reality* is this.'

---

15. Simone Weil, *Cahiers*, vol. 1 (Paris, 1970), p. 158. [Cf. *The Notebooks of Simone Weil*, trans. Wills, p. 45. Translator's Note.]

16. Cf. Wittgenstein, *Lectures and Conversations on Aesthetics, Psychology, and Religious Belief* (Berkeley, 1966), p. 27; Italian translation, *Lezioni e conversazioni*, p. 96.

But the only effect of such assertions is that we are led to overlook the differences between objects we are identifying, and thus to bypass the obvious fact that it is quite improper to speak of *reality* when we are dealing at best with a case of mere representation, appearance or metaphor. *In reality*, any fact represents itself, rather than 'this' or 'that'. So too the facts that Cantor described, if expressed differently, might have lost much of the fascination that they inspired in some people.

Someone has written that 'Each thing is what it is, and not another thing.' The spirit of coldly rigorous criticism conveyed by such drastic assertions – a spirit in accord with the most orthodox criteria of scientific method – had at least the effect of producing antidotes to the claims advanced by this same spirit. In particular, what Wittgenstein defined as one of man's basest desires in the twentieth century – namely, his superficial curiosity about the latest scientific discoveries – should have been revealed as dubious or inconsistent by the ultimate consequences of science itself.

In this regard, the discovery of the antinomies was a motive that stirred up doubts and second thoughts, but we would be naive to suppose that the crisis of the certitudes that derive from mathematical knowledge remained a phenomenon limited to the workshop of the specialists. Evidence that any appeal to scientific reason was an ambiguous rhetorical expedient of persuasion was furnished by people outside the field of mathematics. In a clearly general sense, but one also relevant to particular sciences, Theodor Adorno noted the inherent incongruity of any philosophical assertion that somehow implied a gesture of persuasion. 'Such gestures,' he wrote, 'are founded on the presupposition of a *universitas litterarum*, an *a priori* agreement between minds able to communicate with each other, and thus on total conformity.' In a spirit close to what characterized dialectic in antiquity, Adorno suggested that the task of dialectic was to eliminate from discourse the 'maniacal fixation' of those who seek to convert their unique and particular point of view into an act of persuasion.[17]

Anyone who has read *Dialektik der Aufklärung* or *Minima moralia*

---

17. Theodor W. Adorno, *Minima moralia* 44; English translation by E. F. N. Jephcott (London, 1974), p. 70; Italian translation (Turin, 1954), pp. 65–7.

knows how Adorno succeeded in linking the profoundest motives that inspire scientific method – which are most recognizable in mathematics – with the alienation that lurks in the least significant acts of life. The process of abstraction effected by mathematics typifies the more general process of separating thought from desire, imagination or emotion – which Adorno considered one of the surest signs of incipient dementia. In this sense, the widespread disdain for Kantian apriorism was emblematic. For the separation of the continuum from time, and the prohibition against *seeing* the intellect's abstractions in the concrete sense of a term, paradoxically entails the atrophy of the very faculties of thought, which are ultimately forced to recognize that 'all knowledge is false, and the only true thing is that for which the question of truth and falsehood cannot even be asked.' This produces the foreseeable effect described in *Dialektik der Aufklärung*. The equation of spirit and world dissolves; and reality, now left to itself, appears increasingly reduced to a 'brutal factual datum'.

Except for the error of those people who could not sacrifice their spiritual life to grasp an invisible essence, the kind of quintessential truth described by Adorno was capable of engendering what Musil's *Man Without Qualities* defined as the 'utopia of the precise life'. The greatest drawback, Musil explained, was not that the scientific approach spontaneously eliminated certain functions and habits of thought. Rather, it was the incompatibility between the truths envisioned by science and the limited forms of ordinary, everyday actions. The 'utopia of precision', Musil wrote, resembles a process that amplifies the elementary relations which the scientist imagines will produce macroscopic results. Since the infinitesimal contains the formal archetype of all the patterns of the sensible world, it acquires in the scientific imagination the utopian dimensions of a reality that is no longer *limited*. By constantly rethinking things in terms of supreme principles abstractly grasped as the culmination of precision, this utopia is capable of falling back into the purely *indefinite*.

To cite Musil's own words: 'Utopia is the experiment in which the possible change of an element may be observed, along with the effects of such a change on the compound phenomenon we call life. If the element under observation is precision itself, one isolates it and allows it to develop, considering it as an intellectual habit and way of life, allowed to exert itself on everything it touches. The logical outcome of this should be a human being full of the paradoxical interplay of exactitude and indefiniteness. He is incorruptibly, deliberately cold, as required by the temperament of precision; but beyond this quality, everything else in him is indefinite. The stable internal conditions guaranteed by a system of morality have little value for a man whose imagination is geared to change. Ultimately, when the demand for the greatest and most exact fulfilment is transferred from the intellectual realm to that of the passions, it becomes evident – as already indicated – that the passions disappear and that in their place arises something like a primordial fire of goodness.

'Such is the utopia of precision. One doesn't know how such a man will spend the day, since he cannot continually be poised in the act of creation and will have sacrificed the domestic hearth fire of limited sensations to some imaginary conflagration.'[18]

Such inquiries into the sources of our alienation, even when transposed to areas far removed from their immediate effects, clearly suggest the opposite of the widespread and familiar belief that science (including mathematics) offers unequalled power for the analysis of concrete facts. In any event, the nature of this oppose view was manifested in the very context of mathematical *facts*, and entered directly into the spotlight as people debated the teachings of the school of Frege, Cantor, Russell and Zermelo. As many writers came to oppose their logicist theses – beginning with Poincaré (1905–6), continuing with Brouwer (1907) and the first proponents of intuitionist theses, and concluding with Gödel's celebrated findings – there emerged unpredictable ideas. They suggested concepts that had formerly seemed outmoded, as well as new formulations that did not provoke those alienated reactions which the repetition of a monolithic truth ultimately inspires.

Some who refused the logicist position and the foundation of the actual infinite on the language of sets were supported by the reappear-

---

18. Musil, *Der Mann ohne Eigenschaften*, II.61; Italian translation, pp. 237–8. [English version, *The Man Without Qualities*, trans. Wilkins, pp. 265–6. Translator's Note.]

ance of the ancient Platonic notions of the relationship between the intellectual understanding of scientific truths and the language developed to describe them. Mathematicians like Poincaré and Weyl re-established a link with this philosophy, which not even Russell had wished to eliminate. In these men, Giorgio de Santillana wrote, not only did the bare skeleton of the scientific method reappear, but also the sort of philosophical conscience, contemplative capacity and intuitive density that one usually attributes to more traditional thought.[19]

Pierre Boutroux, an interpreter of Platonic ideas, observed that 'the Platonists establish a profound separation between "discourse" and "intelligence", and between written science, which is a didactic exposition of a known truth, and the "conception" of scientific truths, which is a direct product of exercising our intuitive faculty on the world of ideal notions.'[20] Another excerpt from Boutroux's *L'idéal scientifique des mathématiciens* appeared in several articles that discussed the debate between empiricists and idealists, and between intuitionists and formalists: 'The mathematical fact is independent of the logical or algebraic wrapping by which we seek to represent it: in effect, the idea that we have of it is richer and fuller than all the definitions we can give it, and than all the forms or combinations of signs or propositions with which it is possible to express it. The expression of a mathematical fact is arbitrary and conventional. By contrast, the fact in itself, meaning the truth that it contains, impresses itself on our spirit beyond any convention. Hence, we could not account for the development of mathematical theories if we chose to regard algebraic formulas and logical combinations as the very objects of mathematical research. On the contrary, all the characters of these theories can be readily explained if we admit that algebra and logical propositions are merely the language into which we translate a set of notions and objective facts.'[21]

---

19. G. de Santillana, *Reflections on Men and Ideas* (Cambridge, MA, 1970), pp. 188–9.
20. See A. Dresden, 'Brouwer's Contributions to the Foundations of Mathematics', *Bulletin of the American Mathematical Society* 30 (1924): 25.
21. See Rolin Wavre, 'Y-a-t-il une crise des mathématiques?', *Revue de Métaphysique et de Morale* 31 (1924): 439.

Perhaps the wisest observation can be found in Simone Weil's *Note-books*. 'Only mathematics,' she writes, 'lets us touch the limits of our intelligence. For we may always believe that an experiment is inexplicable because we don't possess all the data. But there, by contrast, we possess all the data, combined in the full light of the demonstration, and yet we still don't comprehend.'[22]

This incomprehension, which Plato defined in his seventh letter as the knowledge beyond names and images, makes it admissible to grant the existence of contradictions, errors and misunderstandings in an exact science like mathematics. Indeed, Poincaré could assert that 'men don't understand each other, since they don't speak the same language, and since there are languages that cannot be learned.'[23]

Simone Weil suggested that this very fact might favour the rebirth of a dialectic in the field of mathematics – a dialectic perhaps sharing Zeno's 'opening' towards the unfathomable, and in which the paradoxical element would be corroborated by the apparent rigour and formal precision found in theses and proofs. 'If contradiction is what pulls, draws the soul towards the light, contemplation of the first principles (hypotheses) of geometry and kindred sciences should be a contemplation of their contradictions.'[24]

What were the mathematical facts which focused criticism on Russell's logicism and Cantor's language of sets? Before the intuitionist theses were perfected, mathematicians like Lebesgue, Baire and Borel appeared on the scene. After 1905 they formulated their objections to the idea of the actual infinite, to the transfinite, and to the use of Zermelo's axiom. Questions of principle regarding the reliability of the new concepts were compounded by the urgent necessity to correct some gaps in the logical and deductive apparatus of the *Principia Mathematica*: for example, the definability of the upper and lower limits in sets of real numbers. Contrary to the expectation that a satisfactory theory of real numbers would have respected, within the

---

22. Simone Weil, *Cahiers*, vol. 3 (Paris, 1974), p. 141. [*The Notebooks of Simone Weil*, trans. Wills, p. 511. Translator's Note.]
23. See R. Wavre, 'Y-a-t-il une crise?', p. 435.
24. Simone Weil, *Cahiers*, vol. 1 (Paris, 1970), p. 144. [*The Notebooks of Simone Weil*, trans. Wills, p. 34. Translator's Note.]

hierarchy of types these upper and lower limits should have been found in an 'order' that was superior compared to the values assumed by the variable in the set for which they constituted limits. As a remedy, Russell had proposed the axiom of reducibility; but Weyl refuted this, denying the axiom's legitimacy and challenging as erroneous the most refined and approved techniques of analysis. The solutions envisioned by Weyl called for a provisional examination of the need to reduce drastically the notions and methods that had been introduced and consolidated in mathematics even before Cantor. The Aristotelian principle of the excluded third began to be questioned; and to keep its applicability intact, it was proposed that the mathematical continuum be restricted to the consideration of those subsets of numbers concerning which it actually made sense to ask questions that had once been thought legitimate for any class of numbers.

Hilbert undertook to save the integrity of mathematics from the antinomies by resorting to the axiomatic method and to logical calculus, but he himself remained convinced that the use of logical deduction had to admit a further extralogical condition that could be intuited as immediate experience prior to any thought activity. This allied him with Kant, and led him to surmise that the projects of Frege and Russell would fail to achieve their goal.[25] While in agreement with Brouwer, he still admitted that the powers of intuitive thought cannot attain the transfinite, and that the mathematical theorems implying this category cannot therefore be justified as eventual truths based on concrete evidence. His programme envisioned comprehending the use of the transfinite from a 'finitist' point of view. Or in a stricter sense not entirely adequate as regards the goal of effective comprehension, it at least envisioned removing the transfinite apparatus from the demonstration of formulas that did not contain symbols implying it.[26]

In any case, despite his strong resolve to preserve intact what he defined as 'Cantor's new paradise', Hilbert showed an acute awareness

---

25. See David Hilbert, 'Über das Unendliche', *Mathematische Annalen* 95 (1925): 161–90; Italian translation in Carlo Cellucci, *La filosofia della matematica* (Bari, 1967), pp. 161–83.

26. See Georg Kreisel, *Hilbert's Program*; Italian translation in Cellucci, *La filosofia della matematica*, p. 191.

of the difficulties implicit in the use of the infinite, and came to recognize that the actual infinite used in deductive methods was a mere illusion that could only be remedied by a substitution using rigidly finite procedures capable of arriving at the same results. Sensing the difficulty of completely freeing certain primary concepts from intuition and linguistic inexpressibility, Hilbert finally centred his research on the problem of *coherence* and formalized mathematics in such a way that it resembled a game of symbols without significance, in which the conventional choice of rules could guarantee the indisputability of the results.

The attempt to grasp the infinite in its totality by defining a completeness that left nothing outside of itself was destined to prove illusory. In 1931, Gödel showed that mathematics kept on revealing *openings*, or references to something *other* than what could be expressed by a formal system like Hilbert's. In other words, symbolic mathematics was not able – as formalism had promised – to express a closed and exhaustive world of signs, a *complete* formal system.

For any (sufficiently powerful) formal system of mathematics, Gödel indicated two inevitable consequences. (1) There exist relatively elementary and intuitively true propositions which cannot be deduced in the formalism of the system. (2) The statement that expresses the coherence of the system cannot itself be deduced within its own formalism, in the sense that an attempt at deduction would lead to the absurdity of a relation such as $1 \neq 1$.

Weyl commented that 'it is surprising that a construction created by the mind itself, the simplest and most diaphanous thing for the constructive mind, assumes a similar aspect of obscurity and deficiency when viewed from [Hilbert's] axiomatic angle.'[27] Gerhard Gentzen's later proof of the coherence of mathematics turned out to be at odds with Hilbert's programme, and implied a fatal penetration into the transfinite. Weyl called it a 'Pyrrhic victory'.[28]

But the most interesting aspect of the question concerned the idea

---

27. Hermann Weyl, *Philosophy of Mathematics and Natural Science* (Princeton, 1949), p. 220.
28. Ibid.

that inspired Gödel's proof techniques. Their paradoxical intent was evident from their recourse to the diagonal method with which Cantor had shown that the continuum was not countable, and with which Richard had in 1905 illustrated his antinomy.[29] A mechanization of the process that would later illustrate this antinomy was not unknown in the literary and metaphysical tradition. Swift had described it in *Gulliver's Travels* (III, 5) in connection with the 'learning projector' used in the Grand Academy of Lagado; and Raymond Lull employed it in his *Ars magna*.[30]

Gödel's proof could well be considered the contribution of genius to the art of paradox, revealing the incomparable value of *impossibility*, of an *obstacle*, as a vehicle for understanding the *absence* of the infinite-as-totality in this world. The Greeks' intuition of the irreplaceable significance of this absence, and of the absurdity of an explicit revelation of the infinite in the ranks of ordinary appearance (even one linked to a mathematical representation), finally found evident confirmation in the conclusions of certain theorems.

There were other findings: Skolem's paradox and the consequent relativization of the concept of class, the recognition of the irremediable imprecision of the axiomatic instrument, and the refoundation of semantics by means of a proposal to distinguish between the object-languages and the meta-languages in the ranks of an indefinite hier-

---

29. A version of Richard's paradox is described as follows by Elliott Mendelson, *Introduction to Mathematical Logic* (New York, 1964), p. 3. 'Some phrases of the English language denote real numbers, e.g., "the ratio between the circumference and diameter of a circle" denotes the number π. All phrases of the English language can be enumerated in a standard way: order all phrases having k letters lexicographically (as in a dictionary); and then place all phrases with k letters before all phrases with a larger number of letters. Hence, all phrases of the English language denoting real numbers can be enumerated merely by omitting all other phrases in the given standard enumeration. Call the nth real number in this numeration the nth Richard number. Consider the phrase: "the real number whose nth decimal place is 1 if the nth decimal place of the nth Richard number is not 1, and whose nth decimal place is 2 if the nth decimal place of the nth Richard number is 1." This phrase defines a Richard number, say the kth Richard number; but, by its definition, it differs from the kth Richard number in the kth decimal place.' The explicit use of a *denotation* classifies the antinomy among the semantic paradoxes, next to the celebrated paradox of the Liar: 'I am lying.'
30. See H. Weyl, *Philosophy of Mathematics and Natural Science*, p. 224.

archy. All of them were further confirmations and corroborations of the new perspective imposed by Gödel's theorem.

In 1932, Weyl condensed this finding in a neat summary: the infinite is intuitively accessible as an indefinitely open field of possibility, and in this respect would seem analogous to a series of numbers that can be extended unlimitedly.[31] Yet completeness, the so-called actual infinite, lies beyond our reach. Nevertheless, the exigencies of totality impel the mind to imagine the infinite, by means of symbolic constructions, as a closed entity. Weyl further suggested that the primary philosophical interest of mathematics, as well as of physics, should consist in attaining the intrinsic solidity of these symbolic constructions; and the research of Weierstrass, Cantor, Frege, Russell and Hilbert had also worked towards this goal.

Weyl's frequent allusions to Pythagoras, Bruno, Nicholas of Cusa, and to the celebrated maxim of Anaxagoras ('Of the small there is no minimum, but always something smaller'), were bewildering to anyone trying to follow closely the evolution of the ideas that culminated in his book *The Open World* (1934). Two years later, Eric Temple Bell confessed that he had been deeply impressed to see Nicholas of Cusa 'raised from the dead'. When he considered the authority of Weyl in discussing the philosophy of the infinite, he ultimately accepted as at least defensible the thesis that the modern philosophy of the infinite had descended from medieval theology.[32]

31. H. Weyl, *The Open World: Three Lectures on the Metaphysical Implications of Science* (New Haven, 1932), p. 83.
32. Eric Temple Bell, 'Finite or Infinite?', *Philosophy of Science* 1 (1934): 31.

13

Open Infinity

Despite his expressed opposition to Hegelian dialectic, Karl Popper posited a *negative* concept, such as the susceptibility to falsification, as a criterion for the *essential* definition of a scientific theory. A theory is *defined* as scientific if it can be refuted; and it is thus evident that at the moment of its refutation we would touch, as he puts it, 'the reality', and so would attain the third and most complete stage of penetration of the theory. (In the first stage, we 'grasp' its general sense; in the second, we can repeat and communicate it; in the third, the theory can be refuted.)

Although his intention was quite different, Hegel had posited as an *essential* element of the finite the moment of its negativity, of its non-being, which could be exemplified in any coherent scientific theory. 'When we say of things that they are finite, we mean that their nature and their being is constituted by non-being.' It is this very non-being, he explains, which reveals the presence of the infinite in the world, and which thus interpreted translates privation (Aristotle's *steresis*) into something *affirmative*. Hegel writes: 'If reality is taken in its determinateness, then, since it essentially contains the moment of negativeness, the sum-total of all realities equally becomes the sum-total of all negations and of all contradictions; it becomes, for instance, absolute power, which holds absorbed everything determinate, but since it exists only insofar as it has opposed to it something not yet subjected to it, any thought which extends it to perfect and unbounded power leads it back to abstract nothing. This real in all reals, or being in determinate being, which is to express the concept of God, is nothing else than abstract being, and identical with nothing.'[1]

Hegel's attempt to grasp the negativity which cloaks the infinite in its contact with the world can sometimes be detected within the interstices of the many theories of knowledge which regarded Hegelian dialectic as a pointlessly constrictive schema.

For example, if we consider the work of Imre Lakatos, who in many ways resembles Popper, we find many examples that cause us to appreciate one of Hegel's profoundest goals: to unify the two apparently irreconcilable faces of the infinite – pure affirmation and pure negation – within the events and objects of the world. In Feyerabend's words, 'we must praise Lakatos for the excellent use he has made of his Hegelian education. On the other hand, it is perhaps necessary to criticize him for not revealing more directly the source of his inspiration, and for giving the impression of turning to a school of thought that is much less comprehensive and far more mechanical.'[2] Even the pragmatic common sense of Popper's philosophy of science is silent about the deeper significance of Hegelian dialectic, which he views as inadequate to explain the evolutionary changes in science. Lakatos' 'Proofs and Refutations' clearly shows how refutations and counter-examples promote *definitions*, progressively refine their sense and precision, and even prompt the mysterious emergence of the formative idea, that is, of the perspective from which we contemplate an object according to criteria of authentic originality.

Lakatos seeks to deprive mathematical proof of its apparently unidirectional nature. Outside the boundaries of purely scientific endeavour, one might add that this is a case of liberating the proof procedure from *desire*, that is, from the illusory pretence of finding what one seeks. If someone is in a hurry to arrive at a conclusion, Lakatos replies: 'You are interested only in proofs which "prove" what they have set out to prove. I am interested in proofs even if they do not accomplish their

---

1. G. W. F. Hegel, *Wissenschaft der Logik* I.1.2; Italian translation, *Scienza della logica* (Bari, 1974), p. 130. [English version, *Science of Logic*, trans. W. H. Johnston and L. G. Struthers, 2 vols. (London, 1929) 1:125. Translator's Note.]
2. Paul K. Feyerabend, *Against Method*; Italian translation, *Contro il metodo* (Milan, 1973), p. 160, n. 91.

intended task. Columbus did not reach India, but he discovered something quite interesting.'[3]

In this phase, an investigator finds himself in the company of a simulacrum of the *void*, which one might comfortably define as the rejection of all scientific idolatry. Naturally, the void can be inhabited by 'monsters' (exceptions, counterexamples, conceptual deformations) and can leave room for the belief that 'if we want to learn about anything really deep, we have to study it not in its "normal", regular, usual form but in its critical state, in fever, in passion.' From this, there derives a bit of extreme and ambiguous advice: 'If you want to know the normal healthy body, study it when it is abnormal, when it is ill.'[4]

Such a survey of irregularities and anomalies constitutes the negative moment in every experiment which even the apodeictic limits of a definition or a mathematical proof may not be able to illuminate. Lakatos even deprives the proof of its apparent character of infallibility whenever it seeks to present itself as proof of a conclusive 'truth'. He cites G. H. Hardy, who writes: 'There is strictly speaking no such thing as mathematical proof; we can, in the last analysis, do nothing but point . . . proofs are what Littlewood and I call *gas*, rhetorical flourishes designed to affect psychology, pictures on the board in the lecture, devices to stimulate the imagination of pupils.' As Lakatos notes, György Polya has also observed that proofs, even incomplete ones, do nothing more than establish 'connections between mathematical facts, and this helps retain them in our memory: proofs produce a system of mnemonics'.[5] Wittgenstein too advised that, to grasp the meaning of a thesis, one should mainly follow the sense of its *proof*.

This explains, in part, the fact that the twentieth-century recovery of infinity as pure potential was furthered by the growth of a radical critique of the act of persuasion, which involved the apodeictic nature of mathematical proof. The 'gesture' of persuasion, which both Adorno and Wittgenstein were able to unmask with decisive arguments that literally stripped bare the organizing structure of knowledge, re-

---

3. Imre Lakatos, 'Proofs and Refutations', *British Journal for the Philosophy of Science* 14 (1963): 15.
4. Ibid., p. 25.
5. Ibid., pp. 125–6.

acquired the ancient ambiguity implied by the Greek verb *peitho*, which Homer uses in *Odyssey* 2.106 in the dual meaning 'to convince' and 'to deceive or delude'.

In a broader sense, one could even say that Husserl and Heidegger were right. In his *Crisis of the European Sciences*, Husserl implied that an ideal mathematics managed by an all-embracing Method had, since Galileo and Descartes, irresistibly tended to be transformed into a reifying technique, a descriptive procedure completely separate from the world of subjectivity, and therefore one that led to alienation and failure. The intrinsic dubiousness of any mathematical representation of the infinite elaborated in the spirit of Descartes, Leibniz or even Cantor could then be explained by verifying a simple fact. If we denote the infinite by a sign, and immerse it in a 'universal mathematics', which we believe can rationally explain a reality separate from its subject, then we create a division 'objectivizing' the truth in forms that are absolute, stable and uniquely significant; and thus we tend to confuse a representation with the invisible being that it merely announces but also conceals in its function as symbol.

If it is true, as Norbert Wiener writes, that 'the thought of every period is reflected in its technique,' we could also detect in some aspects of Heidegger's thought close ties with the crisis in the foundations of mathematics.[6] The discovery of the antinomies plays an implicitly important role in the opening of his *Sein und Zeit*, where he writes that 'the level of a science is measured by the amplitude within which it is able to accommodate the crisis of its fundamental concepts.'[7] Furthermore, the opening caused by the collapse of the certainties and by what Heidegger defined as Being's intrinsic dubiousness (*'Fragwürdigkeit'*) exposes the range of ideas which form the foundation of science as positive research. Hence, each further investigation is automatically forced to face the problem of the *being* of things – a problem different from, but in part analogous to, what Popper means when he writes that refutation approaches reality – and this, we might say, is precisely

---

6. Norbert Wiener, *Cybernetics* (New York, 1948), p. 49; Italian translation, *La cibernetica* (Milan, 1968), p. 65.

7. Martin Heidegger, *Sein und Zeit*; Italian translation, *Essere e tempo* (Turin, 1978), p. 62.

the *being* that Bolzano sought to exclude from scientific language. The representation of the infinite as a mathematical *sign* could be taken as a perfectly emblematic aspect of what Heidegger defined in *Holzwege* as the fundamental trait of the modern world, namely, 'the conquest of the world resolved into images', the aggressive struggle to plan and control all things as definitively stabilized in their objectivity.[8]

Still, as Heidegger writes, the extreme consequence of this will to dominate had to be a confrontation of the *incalculable*, and hence in mathematics the antinomy. The incalculable then became 'the indivisible shadow that spreads over all things when man becomes a *subjectum*, and the world an image'.[9] This shadow was perhaps merely the dimming of the idols. Indeed, thought had transformed into 'idols' what Heidegger considered the complex aggregate of beings, of the things we normally encounter: 'Being is everything of which we speak, of which we think, and with regard to which we behave in one way or another.'[10] How then could one continue to believe that mathematical *images* of the transfinite or the irrational could, in their bare objectivity, offer definitive indications of what the infinite *is*?

Still, it is also true that idolatry, as Simone Weil remarked, is a 'vital necessity' for people in Plato's cave. 'Even with the best of us, it is inevitable that it should set narrow limits for mind and heart.'[11] Naturally, we cannot long resist our horror of the void: 'The void is the supreme plenitude, but man has not the right to know this.'[12]

Heidegger obviously grasped the risk of being suspended in the void when he observed that, by discovering the incalculable, 'the modern world has placed itself in a region that eludes representation'.[13] Man should have learned the truth and guarded it by pure reflection, keeping to an ambiguous territory suspended above that abyss which separates

---

8. Martin Heidegger, *Holzwege*; Italian translation, *Sentieri interrotti* (Florence, 1968), p. 99.

9. Ibid., p. 100.

10. Martin Heidegger, *Sein und Zeit*; Italian translation, *Essere e tempo* (Turin, 1978), p. 59.

11. Simone Weil, *Cahiers*, vol. 2, p. 18. [*Notebooks*, trans. Wills, p. 150. Translator's Note.]

12. Ibid., p. 17. [*Notebooks*, trans. Wills, p. 149.]

13. Heidegger, *Holzwege*; Italian translation, *Sentieri interrotti*, p. 101.

Being from being. Man belongs to being 'yet remains a stranger in Being'. This fatal attraction of entity and its reification could at length take shape, as Adorno wrote with rhetorical ruthlessness, as a new 'objectivity of quasi-superior rank'.[14] Yet this fatal striving of our oscillating thought towards an inexpressible goal is irresistibly viewed as the *positive* content of the new philosophizing. To be sure, Heidegger wrote, 'the understanding of Being, despite all its rank, is dark, confused, covered over and concealed,' but for this very reason 'it must be illuminated, disentangled and ripped away from concealment.'[15] The 'new questioning' that automatically sprang from the perceived 'dubiousness' of the entity created an *open* space, but it was understood that this space should initiate 'a new space that includes and intersects everything'.[16] The seductive power of historical contingency might incautiously have translated this new positivity into an event that was precise and once again (paradoxically) *calculable*.

The 'misfortune' that struck Heidegger personally in his contact with the 'vulgarity' of history has been recounted by Karl Löwith. In Heidegger's speech, delivered when he became rector at Freiburg University in 1933, Löwith saw 'that equivocal confusion of effective history and the authentic occurrence of Being', which is also the crucial point of attraction of the abyss and of the infinite.[17] It is no coincidence that the emblematic character in Hermann Broch's *The Guiltless*, who adopts the name Zacharias, cites the German people as the custodian and interpreter of the mystery of the Infinite. His discourse is consummate, and it is also an indirect comment that justifies the misunderstanding of history which we infer from Heidegger's *Holzwege*. Heidegger implies that Being withdraws at the precise instant when it discovers itself in its being. This withdrawal produces a fatal error which belongs to the essence of truth; hence, the becoming of history

---

14. Theodor Adorno, *Dialettica negativa* (Turin, 1970), p. 98. [English version, *Negative Dialectics*, trans. E. B. Ashton (New York, 1973), p. 109. Translator's Note.]

15. Martin Heidegger, *Einführung in die Metaphysik*; Italian translation, *Introduzione alla metafisica* (Milan, 1968–79), p. 93. [English version, *Introduction to Metaphysics*, trans. Gregory Fried and Richard Polt (New Haven, 2000), p. 87. Translator's Note.]

16. Ibid., p. 40.

17. Karl Löwith, *Saggi su Heidegger* (Turin, 1966), p. 56.

is by its essence the realm of distortion, of the illusion of a madman ('*Irrer*').[18] The face of truth must then be death and annihilation, and whoever justly interprets it respects the inexorable historical discovery of Being, and obeys a fatal injunction that admits of no recourse or appeal.

Zacharias approaches the problem in a prophetic way, using the emphatic language of a person who goes to his death without experiencing its cathartic and regenerative effects: '. . . time and again we must sink into evil in order to raise ourselves and the rest of the world to a state of higher perfection, and time and again in our hands injustice has been transformed into justice, to our own surprise. Because we are the nation of the infinite and hence of death, while the other peoples are bogged down in the finite, in shopkeeping and money-grubbing, confined to the measurable world because they know only life and not death and consequently, though they may seem to rise so easily and so far above themselves, they are unable to break through the finite. It is incumbent on us, for their own salvation, to subject them to the punishment of a death-pregnant infinity. A hard, a colossal course of instruction, forsooth! A hard course to follow, and an even harder one to give, all the harder because not only the dignity of the judge, but also the indignity of the executioner has been imposed upon us, the teachers. For in the infinite all exist side by side, dignity and indignity, sanctity and the need for salvation, goodwill and ill will, hence the curse and the blessing that are our destiny, the dual role in which we become objects of terror to ourselves and others: every shot we are compelled to fire against them strikes our own hearts as well, every punishment we are obliged to mete out to them is our own punishment as well. Our mission as world teacher is a curse and a blessing, and nevertheless we have taken it upon ourselves for the sake of truth which is in infinity and therefore in ourselves: as Germans we have taken it upon ourselves; we did not shrink back from it, because we knew that we, alone of all peoples, are free from hypocrisy.'[19]

---

18. Ibid., p. 58.

19. Hermann Broch, *Die Schuldlosen*; Italian translation, *Gli incolpevoli* (Turin, 1963), p. 141. [English version, *The Guiltless*, trans. Ralph Manheim (Boston, 1974), pp. 143–4. Translator's Note.]

But why does Infinity invoke death? Because it doesn't tolerate the *fiction* of itself, we might perhaps reply with Zacharias. When man seeks infinity, he finds only fictions and the content of gratifying 'reality', which supposedly is furnished by the facts, now transformed into idols by common instinct or by sensationalist dogma, and he thus ultimately creates a deception based on bad faith, and a real distortion. Although Zacharias doesn't see it, this distortion comes from a simple fact: the 'reality' of the world continues to present itself, in the eyes of those who seek to transcend it, with the credentials of an irresistible and alluring concreteness.

It is thus evident, on the other hand, that the amorous involvement of Zacharias with Philippine aspires to suicide. Indeed, more than death Zacharias fears what tormented Hoffmann's Medard: 'The basest material instincts, disguised as mystical raptures, are unleashed in us, promising us here on earth the satisfaction of our most exalted dreams; our unconscious passion is deceived by this, and our aspiration to holy and otherworldly things is shattered in the ineffable and unexperienced joy of the senses.'[20]

As Broch commented, the path of truth is thus infinitely narrow. It separates two worlds and exposes those who travel it to the moral risk of an irreversible fall. For Broch, the ultimate principles of poetic art and scientific thought finally meet on this path, especially after the collapse of the apodeictic positions of nineteenth-century positivism, caused by the discovery of the antinomies. The unity of science (philosophy) and poetry, he explains, is achieved only in the *foundations* of knowledge, and cannot rely on the parallelisms and facile analogies that amaze the observer at first glance.

But how does our intuition of the infinite act in science and particularly in mathematics? Essentially, the mystery is this: 'Without the previous knowledge of an unknown quantity no problem could be stated, and mathematics is no exception. It is this foreknowledge of the infinite and of the continuum which pushed mathematics towards the increasingly complex construction of the infinite multiplicity of

---

20. E. T. A. Hoffmann, *Gli elisir del diavolo*, in *Romanzi e racconti*, vol. 1 (Turin, 1969), p. 456.

beings . . .'[21] For Broch, this foreknowledge is located in a sort of gnoseological unconscious to which we owe every explanation of the rational sphere. From it is born the intuition which allows us to attribute precision and truth to the ascending movement of concepts from natural number to the transfinite, which allows no closure whatsoever.

Thus, the meeting point of the various activities of the spirit, which culminates in our foreknowledge of the infinite, is centred in complete secrecy, in the gnoseological unconscious. This point is the source of not only the concepts of mathematics, but of all our more or less conscious intuitions of the 'combinations of things and properties of the world' which enter the sphere of ordinary action. 'Thanks to this knowledge alone,' Broch writes, 'man is able to find his way "intuitively" in daily life, to choose "instinctively" the "just" action among all possible actions, and to assume on occasion (not always) a "rational" attitude even when he gropes forward like a sleepwalker.'[22]

We might say that man is all the more ingenious for holding his world together by making it correspond to an ideal network, sufficiently tight-knit and exhaustive, formed by the intersections of events – which is why Moosbrugger is the height of genius. But the best expression of this interlacing of possible (and impossible) combinations is obviously furnished by mathematics. Only there are problems described, apparently without gaps and at least implicitly resolved with a precision that no other procedure allows. Mathematics has a further advantage, which struck Broch in particular: it is capable of building its *own empirical* sphere of notions and concepts, according to a logic close to intuitionist logic, by means of *effective* constructions that defy contradiction. The more complex these constructions become, the easier it is to recognize that 'the relations of the empirical world are, as it were, a sub-quantity of mathematical relations.'[23] And the concepts of mathematics consequently become an apparatus of ideal units to which all subjective experience can be referred.

---

21. Hermann Broch, *Azione e conoscenza* (Milan, 1965), p. 180.
22. Ibid., pp. 181–2.
23. Ibid., p. 178.

What would happen if we pushed this procedure to the extreme? We would arrive, Broch says, at a 'nirvana of precision, realizable – if at all – by means of the mathematical instruments worked out by Western science'.[24] But this nirvana of precision is a utopia, in the very same sense that Musil intended when he wrote about the 'utopia of the precise life'. Attaining it would mean annihilation. This difficulty, which suddenly arises in all its intuitive clarity, thus finds its exact correlative in the mathematical antinomies. Through the 'mathematization of the world' envisaged by Broch, these antinomies cast their shadow over the entire life of man; and the mathematical infinite thus ends by indicating, in that margin of the unknowable that it clearly requires, entire ranges of impossibility and interdiction in the world of empirical action.

But the 'residue of quality', whose linguistic representation is thwarted by the infinite, is also the soul of the symbolic, as Broch intuited.[25] In the end, it is the infinite that makes of a symbol a mechanism that is not significantly *unambiguous*, but simply allusive and polyvalent, and thus liberating.

Friedrich Waismann quotes this entire passage from Hofmannsthal: 'In the limitless detail and particularity of description, the subject matter itself seems to oppress and overwhelm us: but what would come so close enough to us as to hurt us, were we limited to immediate meanings, resolved itself by virtue of the multiplicity of meaning in the words into a magic cloud, and so behind the immediate meaning we divine another which is derived from it. Thus it is that we do not lose sight of the proper and original sense: where, however, this sense was commonplace and mean, it loses its implicit commonplaceness, and often, as we contemplate the word, we hesitate in our perceptive awareness between the particular reality which it symbolizes and a higher reality, and this in a flash leads us up to the great and the sublime.'[26]

---

24. Ibid., p. 183.
25. Ibid., p. 186.
26. Friedrich Waismann, *Analisi linguistica e filosofia* (Rome, 1970), pp. 111–12. [F. Waismann, *The Principles of Linguistic Philosophy*, ed. R. Harré (London, 1965), p. 178. Translator's Note.]

Waismann applied the principle of linguistic 'polysemy' to the most celebrated enigmas about the infinite, including Zeno's paradox, which he viewed as a result of the sign's uncontrolled polyvalence. Furthermore, Waismann introduced the idea of *action* into the problem of meaning, and made it the factual, if not calculable, basis for a *systematic ambiguity*, for an inevitable opening towards physiological 'causes' or 'motivations' that were not strictly deterministic. There are different ways to consider an action, just as there are different ways to consider a word or linguistic proposition.[27] And, we might add, mathematical language offers the most precise models of this ambiguity.

It is obvious, then, that the mathematical problem of the infinite is automatically projected into the *moral* sphere. This projection is implicitly suggested by Broch, and is clearly prefigured in the works of Nietzsche and Musil.

All this gives us an approximate idea of what some might consider the symbolic power of the mathematical sign. Obviously, this power far surpasses the empiricism in which we find what we might call the merely 'corporeal' life of the symbol. Yet this very empiricism, in which all mathematical activity properly consists, has an irresistible tendency to manifest itself as an autonomous system, one oriented towards a progressive and endless closing of its own gaps, and clearly inclined to treat foreknowledge of the infinite, or of the continuum, as an unknown quantity foreign to the immanent finality of its own development. In a sense, mathematics tends to assume once again the features that Husserl and Heidegger had more or less explicitly ascribed to it.

At this stage, the potential infinity reclaimed after the invasion of the antinomies seems more open than ever. It revives the 'syncategorematic' sense exactly as it was defined in medieval thought, that is, something that proceeds in the finite, moving by simple changes of perspective and paradigms or by gradual increments without any final goal whatsoever. This goal does not exist for two distinct reasons. First, we are dealing with an *open* sequence. And second, the final cause of our research activity is linked to our foreknowledge of the absolute, which is usually dismissed as being an unknown or a pseudo-problem. Thus, the re-

27. Ibid., p. 127.

evaluation of the infinite as pure negativity, which was caused by the discovery of the paradoxes, fades into nothing. This happens not because we forget the impossibility of the actual infinite, but because we ultimately face a kind of knowledge centred on what is operative and effectual: a *positive* knowledge that generally forgets the disturbances and emotional effects caused by the antinomies.

In mathematics, this 'constructive' and pragmatic aspect, which crept in after the crisis in foundations, can already be discerned in certain statements of von Neumann and Gödel. In 1954, von Neumann proposed that the *use* of mathematical notions that had not been completely filtered through a rigorous critique could be considered legitimate.[28] Gödel observed that the paradoxes of set theory did not justify our adopting a *negative* attitude towards this theory. The problem was undeniably serious, but it was more closely related to logic and epistemology than to mathematics. Gödel also suggested basing a criterion of the truth of axioms on their fecundity, both in mathematics and in physics. We can decide that the 'success' of an axiom gives us a measure of its 'truth', he wrote, rather than any presumably intrinsic 'necessity'.[29] In a recent article, Gregory J. Chaitin observed that Gödel's theorems, after an initially disruptive effect, had almost no influence on the *everyday activities* of mathematicians.[30] (According to von Neumann, we should show all due respect to such *everyday activities*: 'I think it is extremely instructive to watch the role of science in everyday life, and to note how in this area the principle of *laissez faire* has led to strange and wonderful results.'[31]) Chaitin's remarkable idea is worthy of our attention. He examines the question not so much of a theorem's *provability* in a formal system, as of the degree of *complexity* of its proof. As Chaitin and Kolmogorov have shown, this complexity is also a measure of the degree of *information* contained in the theorem being proved.

---

28. See John von Neumann, *Collected Works*, ed. A. H. Taub, vol. 6 (New York, 1963), pp. 480–81.

29. See Kurt Gödel, 'What is Cantor's Continuum Problem?'; Italian translation in Cellucci, *La filosofia della matematica* (Bari, 1967), pp. 113–36.

30. Gregory J. Chaitin, 'Information-Theoretic Limitations of Formal Systems', *Journal of the Association of Computer Machinery* 21 (1974): 403–24.

31. John von Neumann, *Collected Works*, vol. 6, p. 490.

The mathematician, Chaitin explains, has two options. The first is to accept the conjectures that have been formulated on the basis of their empirical corroboration, exactly as experimental scientists do. The second is to return to more general principles from which to proceed deductively. The second option, he notes, has a higher price: the often enormous complexity of its proofs. It is no coincidence that the well-known findings of Presburger on the decidability of the additive theory of integers suffer from a drawback. As Fischer and Rabin later showed more clearly, every *algorithm* constructed for deducing the theorems of that theory encounters propositions that can be proved only in an excessive number of passages. This is a case of what is usually called an *intractable* problem. What can we do, then? By a singular circumstance, the epistemological obstacle of Gödel's theorems unites at this point with the proofs of intractability and is transformed into a wholly *positive* moment in mathematical research.

This transition is made evident in the clearest way by one of Chaitin's observations. 'Gödel's theorem,' he explains, 'does not mean that mathematicians must give up hope of understanding the properties of the natural numbers; it merely means that one may have to adopt new axioms as one seeks to order and interrelate, to organize and comprehend, ever more extensive mathematical observations.' Using complexity as a measure of the information contained in a system of axioms would then make available an *open* set of hypotheses in the light of which a mathematician could work roughly like a physicist. 'The mathematician,' Chaitin adds, 'shouldn't be more upset than the physicist when he needs to assume a new axiom; nor should he be too horrified when an axiom must be abandoned because it is found that it contradicts previously existing theory, or because it predicts properties of the natural numbers that are not corroborated empirically.' In explicit terms, the conclusion is formulated as follows: 'There may be theoretical justification for regarding number theory somewhat more like a dynamic empirical science than as a closed body of theory.'[32]

---

32. The conclusions reached by Chaitin using the notion of complexity are basically the same as those envisaged by Imre Lakatos: see his 'Mathematics, Science, and Epistemology', in vol. 2 of his *Philosophical Papers*, ed. J. Wottall and G. Currie (Cambridge, 1978). Lakatos cites Russell, Fraenkel, Carnap, Curry and Mostowski as

For the reasons cited, the notion of complexity could thus be placed at the top of the mathematical edifice. It would not only help to define and measure mathematical principles – or better, its axioms – but also act as judge and arbiter in choosing possible ways of developing proofs and illustrating results. If we ask how many bits of information-complexity form a set of axioms, and how complex the proof of a theorem or the realization of a verification is, we automatically have new criteria for deciding whether a widely corroborated conjecture or a new axiom has the right to be included in the theory. In this perspective, we can, for example, visualize Riemann's celebrated hypothesis, which is more or less conveniently assumed to be axiomatic whenever we pose the problem of verifying whether a natural number is a prime number.[33]

If we wished to hazard a tentative parallel between a scientific choice and an ethical or even metaphysical one, we might feel the influence of an observation by Alexander Piatigorsky on Buddhist thought: 'The greatest metaphysical problem in Buddhist philosophy is the problem of the complexity of everything; to be more precise, it is that every phenomenon is complex in the sense that any thing can subsist as a phenomenon only to the extent that it is complex.'[34] From this, we might at least derive an indication of the profound sense (involving all levels of existence) that is acquired by renouncing direct and unambiguous questions like 'What is this or that?' – which in fact Buddhism has generally avoided as misleading, preferring the empirical path of the *fluidity*, *opening* and *being in relation* to phenomena.

Without consulting Buddhism, we could also find inspiration in books like Joël de Rosnay's *Le macroscope* (*The Macroscope*), and in

---

the precursors of a vision of mathematics as an empirical science. Mostowski appeals to Gödel's 'negative' findings in a way that closely recalls Chaitin: 'Gödel's findings and other negative findings,' he writes (p. 27), 'confirm . . . that mathematics . . . is an empirical science, and that its notions and methods have their roots in experience.'

33. See G. L. Miller, 'Riemann's Hypothesis and a Test for Primality', *Journal of Computer and Systems Science* 13 (1976): 300–317; and M. O. Rabin, 'Probabilistic Algorithms', in *Algorithms and Complexity: Recent Results and New Directions*, ed. J. F. Traub (New York, 1976), pp. 21–40.

34. Alexander Piatigorsky, 'La riontologizzazione del pensiero nel Buddhismo', *Conoscenza Religiosa* (1978): 342–66.

features of the systemic approach to understanding infinitely complex structures.[35] Joël de Rosnay does not hesitate to associate such features (often established in a rigidly scientific way) with a proposal for an ethics and, as it were, for an original *Weltanschauung* offering potential references to spheres of wisdom which official science prefers to ignore. (In fact, considerable weight is given to a statement like that of von Neumann, which could easily be interpreted reductively: 'It is truly important that mathematics establish certain standards of truth; and it is truly important that it furnishes a means to establish these standards in a way that is sufficiently independent of everything else': sufficiently independent, as well, of 'emotional' or 'moral' questions.)

In the final analysis, all of this has its roots in mathematics, and in the hasty fashion in which it has been made to conform to a model of science as a general theory of *automata* – a concept which plays an essential role in de Rosnay's book and in the idea of complexity. When we understand the laws governing the function of an infinitely complex system (such as a cell, an organism, a city or an ecosystem), we apply criteria which globally detect the interactions of various components. Such criteria resemble those outlined by John von Neumann in his first description of automatons and their similarity to living organisms. In a formula employed by von Neumann, the automaton could be conceived as a 'reliable organism' (in a sense established by the *statistical* theory of error), based on the reciprocal interaction of singularly *unreliable* components.[36] This is an example of a system involving a large-scale structure and global effects, observable from a distance at which, so to speak, a microscope is useless. Furthermore, the mathematical model on which the 'ideal' and simplified structure of the automaton is based (momentarily divorced from the statistical problem of error) is the one furnished by intuitionist logic, that is, by a constructive logic that is largely immune to the ambiguities of the infinite. Turing, McCullogh, Pitts and von Neumann himself represented

---

35. [Joël de Rosnay, *Le macroscope: vers une vision globale* (Paris, 1975); English version, *The Macroscope: A New World Scientific System*, trans. Robert Edwards (New York, 1979). Translator's Note.]

36. John von Neumann, 'Probabilistic Logics and the Synthesis of Reliable Organisms from Unreliable Components', in *Collected Works*, vol. 5, pp. 329–78.

intuitionist logic in terms of electrical networks, or of idealized nervous systems; and this model became an ulterior, reassuring and potent argument (with an overwhelming factual charge) for the decision to operate as much as possible on the terrain of the *finite*.

In conclusion, we must ponder the words with which John von Neumann described the unsurpassed conceptual *flexibility* of mathematics. Without the obscurities, doubts and uncertainties which attend exactitude, one could scarcely conclude as he does: 'I feel that one of the most important contributions of mathematics to our thinking is that it has demonstrated an enormous flexibility in the formation of concepts, a degree of flexibility to which it is difficult to arrive in a non-mathematical mode.'[37]

---

37. John von Neumann, *Collected Works*, vol. 6, p. 482.

Index

'Among the books in the last few years which I have most often read, reread and thought about' Italo Calvino

The idea of infinity has fascinated us from classical civilization to modern times. For some it represented chaos and terror. For others it was a manifestation of God. Even today it conjures up the image of an endless void, beyond the realms of human imagination. Is there a way to define infinity? How can we describe the incalculable?

Paolo Zellini's masterpiece explores every aspect of infinity, distilling the wisdom of philosophers, artists, mathematicians and theologians over millennia, from Aristotle to St Thomas Aquinas, Bertrand Russell to Robert Musil. What is the difference between true and false infinity – and how can the legend of Sisyphus, for ever doomed to roll a rock up a hill, illustrate this? How might we explain the puzzle of Zeno's paradox? Does 'infinite' mean 'indefinite' – and what is the definition of 'transfinity'?

With grace, lucidity and passion, Zellini unravels the mysteries of this endlessly fascinating concept, and tells the extraordinary story of humanity's quest to solve the unfathomable enigma of the infinite.

U.K. £8.99 U.S.A. $17.00 CAN. $20.00   Philosophy/Popular Science
Cover design: David Pearson; illustration © Image Bank/Getty

read more

ISBN 0-141-00762-1

9 780141 007625

penguin.com